The Challenges and Therapeutic Prospects in Eye Disease

The Challenges and Therapeutic Prospects in Eye Disease

Editor

Chieh-Chih Tsai

MDPI • Basel • Beijing • Wuhan • Barcelona • Belgrade • Manchester • Tokyo • Cluj • Tianjin

Editor
Chieh-Chih Tsai
Department of Ophthalmology
Taipei Veterans General Hospital
Taipei
Taiwan

Editorial Office
MDPI
St. Alban-Anlage 66
4052 Basel, Switzerland

This is a reprint of articles from the Special Issue published online in the open access journal *Journal of Personalized Medicine* (ISSN 2075-4426) (available at: www.mdpi.com/journal/jpm/special_issues/Prospects_Eye_Disease).

For citation purposes, cite each article independently as indicated on the article page online and as indicated below:

LastName, A.A.; LastName, B.B.; LastName, C.C. Article Title. *Journal Name* **Year**, *Volume Number*, Page Range.

ISBN 978-3-0365-8403-4 (Hbk)
ISBN 978-3-0365-8402-7 (PDF)

Contents

Preface to "The Challenges and Therapeutic Prospects in Eye Disease"

A number of key insights in relation to eye disease have been revealed in the recent decade, which has resulted in the development of novel effective targeted therapies, such as teprotumumab for the treatment of thyroid eye disease (also known as Graves' orbitopathy) and intraocular injections of anti-vascular endothelial growth factor (VEGF) agents for many retinal diseases. This Special Issue aims to update the current evidence and challenging topics on the diagnosis and management of ocular diseases.

Chieh-Chih Tsai
Editor

Journal of
Personalized Medicine

Editorial

The Challenges and Therapeutic Prospects in Eye Disease

Chieh-Chih Tsai [1,2]

1 Department of Ophthalmology, Taipei Veterans General Hospital, Taipei 11217, Taiwan; cctsai@vghtpe.gov.tw
2 School of Medicine, National Yang Ming Chiao Tung University, Taipei 11221, Taiwan

A number of key insights into eye disease have been revealed in the past decade, which has resulted in the development of novel, effective, targeted therapies such as teprotumumab for the treatment of thyroid eye disease (also known as Graves' orbitopathy) [1] and intraocular injections of antivascular endothelial growth factor (VEGF) agents for many retinal diseases [2]. This Special Issue aims to update the current evidence regarding challenging topics on the diagnosis and management of ocular diseases. We collected ten original research articles and two literature reviews on recent efforts made toward the discovery of novel findings in different ocular research areas, including orbital, lacrimal, and eyelid disease; glaucoma; retinal disease; myopia; and corneal disease.

Orbital, lacrimal, and eyelid disease

IgG4-related diseases (IgG4-RDs) are an emerging fibro-inflammatory condition that can be characterized by affected organ enlargement, infiltration of lymphocytes and IgG4-positive plasma cells in various tissues and organs, and serum IgG4 level elevation [3]. Serial changes in serum IgG4 levels is a useful adjunct diagnostic method in the assessment of disease activity in patients with IgG4-ROD [4]. Chou et al. further investigated the relationship between the long-term changes in serum IgG4 levels and the clinical course of patients with IgG4-ROD [5]. They found 64% patients with IgG4-ROD had persistently high serum IgG4 levels during long-term follow-up. Importantly, their study disclosed 40% of patients with IgG4-ROD remained in remission despite persistently elevated serum IgG4 levels. Chou et al. suggest these patients can be followed-up without treatment unless disease relapse occurs [5].

Hydrogel scleral buckles were first introduced in the 1980s as an alternative to silicone buckles for the treatment of rhegmatogenous retinal detachment. Hydrogel scleral buckles may absorb tissue fluids and progressively expand over the decades, causing compression of the eyeball, and are commonly misdiagnosed as orbital tumors [6]. The removal of swelling hydrogel scleral buckles is challenging due to their fragile characteristics and fibrotic adhesion to the surrounding ocular tissues. Yang et al. introduced a small excision surgical technique to meticulously remove swelling hydrogel scleral buckles to avoid the risk of severe complications associated with manipulations [7].

Graves' orbitopathy (GO), an increase in orbital fat and/or muscle tissue due to autoimmune inflammation, can cause proptosis, double vision, and/or dysthyroid optic neuropathy (DON). Steiert et al. examined visual parameters dependent on the orbital muscle volume fraction in a surgically treated GO cohort and suggested that the orbital muscle volume factor should be addressed during treatment decisions, while early orbital decompression should be considered particularly in decreased vision with orbital muscle enlargement [8]. In a way that is different to the traditional staged GO management order (decompression and then muscle surgery or lid surgery), Hsieh et al. suggested that orbital decompression combined with strabismus surgery by experienced surgeons can achieve satisfactory outcomes in selected patients, especially for those with symmetry of orbitopathy, relatively simple strabismus and mild proptosis [9]. Choi et al. further supported the use of customized orbital decompression surgery combined with eyelid or strabismus surgery in mild to moderate GO patients, as it had favorable cosmetic results comparable to those achieved through stepwise techniques [10].

Citation: Tsai, C.-C. The Challenges and Therapeutic Prospects in Eye Disease. *J. Pers. Med.* **2023**, *13*, 930. https://doi.org/10.3390/jpm13060930

Received: 30 May 2023
Accepted: 30 May 2023
Published: 31 May 2023

Balloon dacryocystoplasty is a minimally invasive procedure that can achieve lacrimal system patency and is usually performed mainly in cases of congenital nasolacrimal duct obstruction with failed probing [11]. It is known to be more effective than probing alone in older children after 18 months of age with congenital nasolacrimal duct obstruction [12]. It may also be used for the treatment of adult patients with partial nasolacrimal duct obstruction, especially in those who are poor candidates for dacryocystorhinostomy or general anesthesia [13]. Lai et al. found that balloon dacryocystoplasty (DCP) with pushed monocanalicular intubation in adults aged under 65 with complete primary acquired nasolacrimal duct obstruction can achieve a higher success rate than those who underwent balloon dacryocystoplasty alone [14].

Blepharoptosis, the drooping of the upper eyelid, may cause upper visual defects and affect vision. It is important for general practitioners (nonophthalmic physicians) to correctly diagnosis blepharoptosis to assist in decision making for referrals and/or advance work-ups when necessary. Medical artificial intelligence (AI) using machine learning and deep learning has been adopted by various groups to provide effective solutions to challenges facing ophthalmologists and healthcare providers worldwide. Hung et al. developed an AI model with a convolutional neural network (CNN)-based deep learning method to assist in the diagnosis of blepharoptosis, and they demonstrated that the AI model showed better performance than the nonophthalmic physician group in identifying referable blepharoptosis, including true ptosis and pseudoptosis [15].

Glaucoma

Glaucoma is a neurodegenerative disease that affects primarily the retinal ganglion cells (RGCs). Reductions in IOP, achieved with medication, laser, or surgery, has until now been considered as the only effective treatment strategy for slowing down or halting glaucoma progression. However, an even lower IOP does not preclude the possibility of glaucoma progression, as showed by the Collaborative Normal-Tension Glaucoma Study [16]. More and more ophthalmic researchers have tried to develop neuroprotective therapies as a supplement to IOP-lowering treatment. Kuo et al. summarized recent therapeutic advances in IOP-independent neuroprotection research, and discussed the feasibility and hurdles of each potential therapeutic mechanism of various agents in neuroprotection [17]. Nocturnal intraocular pressure (IOP) fluctuations are an important issue in the management of glaucoma. Huang et al. attempted to use a contact lens sensor to monitor the IOP circadian fluctuation change after ripasudil eye drop administration [18]. Although the value of each parameter became lower after the use of ripasudil eye drops, the reduction did not reach statistical significance due to the low baseline IOP in patients with normal tension glaucoma.

Retina

Age-related macular degeneration (AMD), a degenerative disease of the macula, is the most prevalent retinal disease in the Western world, and the neovascular form of AMD may lead to progressive vision loss and cause impairment of the quality of life [19]. Advances in multimodal imaging techniques, fluorescein angiography, indocyanine angiography, optical coherence tomography (OCT), and OCT angiography have been developed for the early diagnosis and follow-up of patients with AMD. Kirkova et al. used OCT-A to provide prognostic markers and new treatment strategies for AMD depending on the naïve neovascular membrane morphologic type [20]. Chen et al. further stated that OCT morphological changes in patients with diabetic macular edema receiving an intravitreal injection of ranibizumab was correlated with a worse structural and visual outcome; thus, serous retinal detachment may be an indicator of an earlier stage of diabetic macular edema, which responds well to the intravitreal injection of ranibizumab [21].

Refractive errors

Increasing studies have revealed that changes in the bioelectrical activity of masticatory and cervical spine muscles may have a primary or secondary effect on the refractive error. [22–24]. Zieliński et al. further showed refractive errors are related to differences in masticatory and neck muscle thickness, and the activity and bioelectrical activity within the temporalis anterior seems to be associated with ocular length, retinal thickness, and choroidal thickness in women with myopia [25]. This may support the finding of interdependence between the stomatognathic system and the organ of vision.

Cornea

Photoactivated chromophore corneal collagen cross-linking (PACK-CXL) is widely known as a minimally invasive therapy for corneal ectasia via stiffening of the cornea tissue by combining ultraviolet A radiation and riboflavin (vitamin B2) through collagen fiber photopolymerization. [26,27] PACK-CXL has also been proposed to be an emerging application to treat infectious keratitis. Infectious keratitis is associated with a risk of visual loss and is one of the leading causes of blindness in developing countries. Infectious keratitis has been conventionally treated with topical broad-spectrum antibiotics or antifungals. PACK-CXL offers a potential, less expensive alternative, acting as a primary or adjunctive treatment for infectious keratitis. Barac et al. carried out a comprehensive review to identify the main diagnostic and prognostic factors involved in therapeutic indications and contraindications of PACK-CXL in infectious keratitis [28].

In conclusion, all articles appearing in this Special Issue provide attractive and current topics that cover a wide range of clinical diagnosis, prognosis, and therapeutic strategies for various ocular diseases. In the future, more investigations will continue to drive our progress toward improving the diagnosis and treatments to meet the challenges and future prospects of eye disease.

Funding: This research received no external funding.

Acknowledgments: I am very grateful to all authors who have provided excellent contributions to this Special Issue. Moreover, I would like to thank the *Journal of Personalized Medicine* for offering me the opportunity to be the Guest Editor of this Special Issue, and in particular, would like to thank Belle Zhang for her help.

Conflicts of Interest: The author declares no conflict of interest.

References

1. Douglas, R.S.; Kahaly, G.J.; Ugradar, S.; Elflein, H.; Ponto, K.A.; Fowler, B.T.; Dailey, R.; Harris, G.J.; Schiffman, J.; Tang, R.; et al. Teprotumumab Efficacy, Safety, and Durability in Longer-Duration Thyroid Eye Disease and Re-treatment: OPTIC-X Study. *Ophthalmology* **2022**, *129*, 438–449. [CrossRef] [PubMed]
2. Pham, B.; Thomas, S.M.; Lillie, E.; Lee, T.; Hamid, J.; Richter, T.; Janoudi, G.; Agarwal, A.; Sharpe, J.P.; Scott, A.; et al. Anti-vascular endothelial growth factor treatment for retinal conditions: A systematic review and meta-analysis. *BMJ Open.* **2019**, *28*, e022031. [CrossRef]
3. Yu, W.K.; Tsai, C.C.; Kao, S.C.; Liu, C.J.L. Immunoglobulin G4-related ophthalmic disease. *Taiwan J. Ophthalmol.* **2018**, *8*, 9–14. [CrossRef] [PubMed]
4. Yu, W.K.; Kao, S.C.; Yang, C.F.; Lee, F.L.; Tsai, C.C. Ocular adnexal IgG4-related disease: Clinical features, outcome, and factors associated with response to systemic steroids. *Jpn. J. Ophthalmol.* **2015**, *59*, 8–13. [CrossRef]
5. Chou, W.Y.; Tsai, C.Y.; Tsai, C.C. Long-Term Follow-Up in IgG4-Related Ophthalmic Disease: Serum IgG4 Levels and Their Clinical Relevance. *J. Pers. Med.* **2022**, *12*, 1963. [CrossRef]
6. Crama, N.; Klevering, B.J. The Removal of Hydrogel Explants: An Analysis of 467 Consecutive Cases. *Ophthalmology* **2016**, *123*, 32–38. [CrossRef] [PubMed]
7. Yang, H.Y.; Yu, W.K.; Tsai, C.C. Management of Delayed Complications of Hydrogel Scleral Buckles. *J. Pers. Med.* **2022**, *12*, 629. [CrossRef] [PubMed]
8. Steiert, C.; Kuechlin, S.; Masalha, W.; Beck, J.; Lagrèze, W.A.; Grauvogel, J. Increased Orbital Muscle Fraction Diagnosed by Semi-Automatic Volumetry: A Risk Factor for Severe Visual Impairment with Excellent Response to Surgical Decompression in Graves' Orbitopathy. *J. Pers. Med.* **2022**, *12*, 937. [CrossRef] [PubMed]
9. Hsieh, M.W.; Hsu, C.K.; Kuo, P.C.; Chang, H.C.; Chen, Y.H.; Chien, K.H. Factors Predicting the Success of Combined Orbital Decompression and Strabismus Surgery in Thyroid-Associated Orbitopathy. *J. Pers. Med.* **2022**, *12*, 186. [CrossRef]
10. Choi, S.W.; Lee, J.Y.; Lew, H. Customized Orbital Decompression Surgery Combined with Eyelid Surgery or Strabismus Surgery in Mild to Moderate Thyroid-associated Ophthalmopathy. *Korean J. Ophthalmol.* **2016**, *30*, 1–9. [CrossRef]
11. Chen, P.L.; Hsiao, C.H. Balloon dacryocystoplasty as the primary treatment in older children with congenital nasolacrimal duct obstruction. *J. AAPOS.* **2005**, *9*, 546–549. [CrossRef]
12. Dericioğlu, V.; Sevik, M.O.; Saçu, S.S.; Eraslan, M.; Çerman, E. Effect of age on primary balloon dacryocystoplasty and probing success in congenital nasolacrimal duct obstruction. *Int. Ophthalmol.* **2022**, *42*, 3547–3554. [CrossRef]

13. Yazici, Z.; Yazici, B.; Parlak, M.; Erturk, H.; Savci, G. Treatment of obstructive epiphora in adults by balloon dacryocystoplasty. *Br. J. Ophthalmol.* **1999**, *83*, 692–696. [CrossRef]
14. Lai, C.C.; Yang, C.J.; Lin, C.C.; Chi, Y.C. Balloon Dacryocystoplasty with Pushed Monocanalicular Intubation as a Primary Management for Primary Acquired Nasolacrimal Duct Obstruction. *J. Pers. Med.* **2023**, *13*, 564. [CrossRef] [PubMed]
15. Hung, J.Y.; Chen, K.W.; Perera, C.; Chiu, H.K.; Hsu, C.R.; Myung, D.; Luo, A.C.; Fuh, C.S.; Liao, S.L.; Kossler, A.L. An Outperforming Artificial Intelligence Model to Identify Referable Blepharoptosis for General Practitioners. *J. Pers. Med.* **2022**, *12*, 283. [CrossRef] [PubMed]
16. Collaborative Normal-Tension Glaucoma Study Group. Comparison of glaucomatous progression between untreated patients with normal-tension glaucoma patients with therapeutically reduced intraocular pressures. *Am. J. Ophthalmol.* **1998**, *126*, 487–497. [CrossRef] [PubMed]
17. Kuo, C.Y.; Liu, C.J.L. Neuroprotection in Glaucoma: Basic Aspects and Clinical Relevance. *J. Pers. Med.* **2022**, *12*, 1884. [CrossRef]
18. Huang, S.K.; Ishii, M.; Mizuki, Y.; Kawagoe, T.; Takeuchi, M.; Nomura, E.; Mizuki, N. Circadian Fluctuation Changes in Intraocular Pressure Measured Using a Contact Lens Sensor in Patients with Glaucoma after the Adjunctive Administration of Ripasudil: A Prospective Study. *J. Pers. Med.* **2023**, *13*, 800. [CrossRef]
19. Heesterbeek, T.J.; Lorés-Motta, L.; Hoyng, C.B.; Lechanteur, Y.T.E.; den Hollander, A.I. Risk factors for progression of age-related macular degeneration. *Ophthalmic. Physiol. Opt.* **2020**, *40*, 140–170. [CrossRef]
20. Kirkova, R.; Murgova, S.; Kirkov, V.; Tanev, I. Personalized Approach in Treatment of Neovascular Age-Related Macular Degeneration. *J. Pers. Med.* **2022**, *12*, 1456. [CrossRef]
21. Chen, N.N.; Lai, C.H.; Lee, C.Y.; Kuo, C.N.; Chen, C.L.; Huang, J.C.; Wu, P.C.; Wu, P.L.; Chen, C.Y. Change of Opticala Coherence Tomography Morphology and Associated Structural Outcome in Diabetic Macular Edema after Ranibizumab Treatment. *J. Pers. Med.* **2022**, *12*, 611. [CrossRef]
22. Monaco, A.; Cattaneo, R.; Spadaro, A.; Giannoni, M.; Di Martino, S.; Gatto, R. Visual Input Effect on EMG Activity of Masticatory and Postural Muscles in Healthy and in Myopic Children. *Eur. J. Paediatr. Dent.* **2006**, *7*, 18–22. [PubMed]
23. Zieliński, G.; Matysik-Woźniak, A.; Rapa, M.; Baszczowski, M.; Ginszt, M.; Zawadka, M.; Szkutnik, J.; Rejdak, R.; Gawda, P. The Influence of Visual Input on Electromyographic Patterns of Masticatory and Cervical Spine Muscles in Subjects with Myopia. *J. Clin. Med.* **2021**, *10*, 5376. [CrossRef]
24. Zieliński, G.; Filipiak, Z.; Ginszt, M.; Matysik-Woźniak, A.; Rejdak, R.; Gawda, P. The Organ of Vision and the Stomatognathic System—Review of Association Studies and Evidence-Based Discussion. *Brain Sci.* **2022**, *12*, 14. [CrossRef]
25. Zieliński, G.; Wójcicki, M.; Rapa, M.; Matysik-Woźniak, A.; Baszczowski, M.; Ginszt, M.; Litko-Rola, M.; Szkutnik, J.; Różyło-Kalinowska, I.; Rejdak, R.; et al. Masticatory Muscle Thickness and Activity Correlates to Eyeball Length, Intraocular Pressure, Retinal and Choroidal Thickness in Healthy Women versus Women with Myopia. *J. Pers. Med.* **2022**, *12*, 626. [CrossRef] [PubMed]
26. Randleman, J.B.; Khandelwal, S.S.; Hafezi, F. Corneal cross-linking. *Surv. Ophthalmol.* **2015**, *60*, 509–523. [CrossRef] [PubMed]
27. Gulias-Cañizo, R.; Benatti, A.; De Wit-Carter, G.; Hernández-Quintela, E.; Sánchez-Huerta, V. Photoactivated Chroqqmophore for Keratitis-Corneal Collagen Cross-Linking (PACK-CXL) Improves Outcomes of Treatment-Resistant Infectious Keratitis. *Clin. Ophthalmol.* **2020**, *14*, 4451–4457. [CrossRef]
28. Barac, I.R.; Artamonov, A.R.; Baltă, G.; Dinu, V.; Mehedințu, C.; Bobircă, A.; Baltă, F.; Barac, D.A. Photoactivated Chromophore Corneal Collagen Cross-Linking for Infectious Keratitis (PACK-CXL)—A Comprehensive Review of Diagnostic and Prognostic Factors Involved in Therapeutic Indications and Contraindications. *J. Pers. Med.* **2022**, *12*, 1907. [CrossRef]

Journal of *Personalized Medicine*

MDPI

Review

Neuroprotection in Glaucoma: Basic Aspects and Clinical Relevance

Che-Yuan Kuo [1] and Catherine Jui-Ling Liu [1,2,*]

1 Department of Ophthalmology, Taipei Veterans General Hospital, Taipei 11217, Taiwan
2 Faculty of Medicine, School of Medicine, National Yang Ming Chiao Tung University, Taipei 11221, Taiwan
* Correspondence: jlliu@vghtpe.gov.tw; Tel.: +886-2-2875-7325

Abstract: Glaucoma is a neurodegenerative disease that affects primarily the retinal ganglion cells (RGCs). Increased intraocular pressure (IOP) is one of the major risk factors for glaucoma. The mainstay of current glaucoma therapy is limited to lowering IOP; however, controlling IOP in certain patients can be futile in slowing disease progression. The understanding of potential biomolecular processes that occur in glaucomatous degeneration allows for the development of glaucoma treatments that modulate the death of RGCs. Neuroprotection is the modification of RGCs and the microenvironment of neurons to promote neuron survival and function. Numerous studies have revealed effective neuroprotection modalities in animal models of glaucoma; nevertheless, clinical translation remains a major challenge. In this review, we select the most clinically relevant treatment strategies, summarize preclinical and clinical data as well as recent therapeutic advances in IOP-independent neuroprotection research, and discuss the feasibility and hurdles of each therapeutic approach based on possible pathogenic mechanisms. We also summarize the potential therapeutic mechanisms of various agents in neuroprotection related to glutamate excitotoxicity.

Keywords: glaucomatous optic neuropathy; glutamate; optic nerve protection

Citation: Kuo, C.-Y.; Liu, C.J.-L. Neuroprotection in Glaucoma: Basic Aspects and Clinical Relevance. *J. Pers. Med.* **2022**, *12*, 1884. https://doi.org/10.3390/jpm12111884

Academic Editors: Juan J. Salazar Corral and Yousif Subhi

Received: 19 September 2022
Accepted: 4 November 2022
Published: 10 November 2022

1. Introduction

Glaucoma is one of the leading causes of irreversible blindness worldwide; it is an optic neuropathy characterized by the progressive loss of the visual field due to the apoptosis of retinal ganglion cells (RGCs) [1]. It is a multifactorial disease with complex pathogenesis that is not yet fully understood (Figure 1).

Intraocular pressure (IOP) is one of the most important risk factors for the development and progression of glaucoma, and IOP-lowering therapy is widely regarded as the only effective treatment strategy for slowing down or halting the deterioration of glaucomatous optic neuropathy. Reductions in IOP can be achieved with medication, laser, or surgery. Most patients with glaucoma, even after laser or surgical treatment, are treated with topical ocular hypotensive medications, which work by reducing the production of aqueous humor or facilitating the trabecular or uveoscleral aqueous outflow. These anti-glaucoma medications include β-adrenergic antagonists; α2-adrenergic agonists; carbonic anhydrase inhibitors; prostaglandin F2a analogs; and, more recently, Rho kinase inhibitors, latanoprostene bunod, and omidenepag isopropyl. In individuals with normal tension or high-tension glaucoma, ocular hypotensive medications are useful in delaying or preventing disease progression [2–4].

Although there are a variety of hypotensive medications and surgical techniques that can efficiently and effectively lower IOP, IOP reduction is sometimes insufficient to prevent glaucoma progression. In the Ocular Hypertension Treatment Study, 4.4% of medicated participants developed glaucoma 5 years after follow-up despite a 22.5% IOP reduction from an average of 24.9 mmHg to 19.3 mmHg [4]. Disease progression also occurred in 45% of treated patients who had a 25% IOP reduction from baseline IOP of

20.6 mmHg in the Early Manifest Glaucoma Trial [3]. Moreover, an even lower IOP does not preclude the possibility of glaucoma progression, as the Collaborative Normal-Tension Glaucoma Study revealed that 12% of treated patients experienced disease progression despite a 30% reduction to an average IOP of 10.6 mmHg during 5.6 years' follow-up [5]. In addition, previous evidence has indicated that glaucoma is primarily an optic neuropathy with the optic nerve head (ONH) being the primary site of the disease [6,7]. As a result, more and more ophthalmic researchers have paid attention to investigating biomolecular mechanisms behind neuronal survival and developing further neuroprotective therapies as a supplement to IOP-lowering treatment.

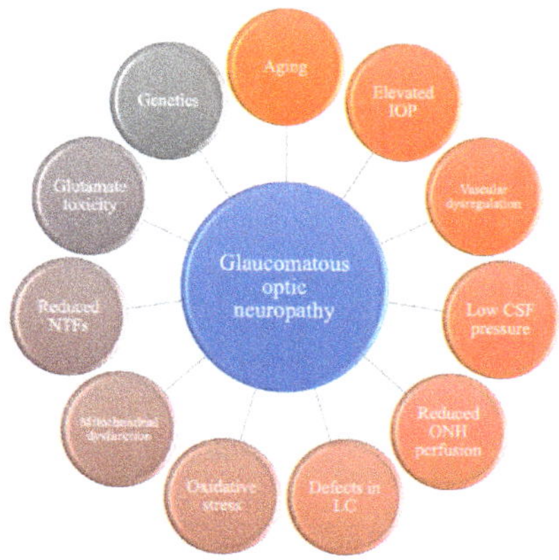

Figure 1. Overview of multifactorial mechanisms contributing to the development of glaucomatous optic neuropathy. CSF, cerebrospinal fluid; IOP, intraocular pressure; LC, lamina cribrosa; NTFs, neurotrophins; ONH, optic nerve head.

Neuroprotection is a therapeutic approach that aims at preserving neural structure and function [8]. In glaucoma, neuroprotection refers to non-IOP-related interventions that can prevent or delay the apoptosis of RGCs independent of IOP [9]. Although it may be difficult to identify a single causative factor for the development of glaucoma, a reasonable approach to tackling glaucomatous optic neuropathy remains targeting possible underlying mechanisms of glaucomatous damage, including the deprivation of neurotrophic factors (NTFs), the formation of reactive oxygen species (ROS), oxidative stress, glutamate excitotoxicity (Figure 2), ischemia, glial activation, and genetic determinants [10]. Therefore, understanding pathogenic factors in glaucoma may further pave the way to the development of more practical neuroprotective methods and subsequent clinical translation.

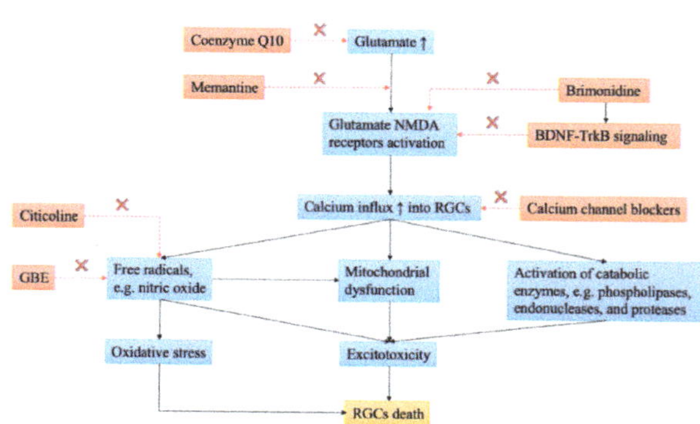

Figure 2. Schematic diagram illustrating the mechanism of glutamate excitotoxicity and summarizing

potential therapeutic mechanisms of several agents in neuroprotection related to glutamate excitotoxicity. The symbol "×" indicates inhibition. BDNF, brain-derived neurotrophic factor; GBE, ginkgo biloba extract; NMDA, N-methyl-D-aspartate; RGCs, retinal ganglion cells; TrkB, tropomyosin receptor kinase B.

2. Neurotrophic Factors

Neurotrophic factors exert a variety of actions by binding to specific receptors and affecting neuron development, survival, and repair via tyrosine kinase signaling [11,12]. Neurotrophic factors are neuroprotective, encouraging axon regeneration and improving neuronal cell function [13]. Because of their promising outcomes in other neurodegenerative disorders of the central nervous system [14–16], NTFs are an appealing therapeutic target to explore in glaucoma.

Several NTFs have been reported to be associated with glaucoma, including nerve growth factor (NGF), brain-derived neurotrophic factor (BDNF), ciliary neurotrophic factor (CNTF), fibroblast growth factor 2 (bFGF), glial-derived neurotrophic factor (GDNF), and neurturin [17–21]. These NTFs have been proven in several glaucoma rodent models to be effective in preventing RGC cell death [17–19,21]. Clinically, topical treatment with NGF for 3 months was shown to enhance optic nerve functions such as visual field, visual acuity, and contrast sensitivity in a small case series of patients with advanced glaucoma [19]. Most recently, a phase 1b clinical trial evaluating the safety and efficacy profiles of recombinant human NGF eyedrops in glaucoma patients for 8 weeks revealed neither major adverse events nor statistically significant short-term neuroenhancement in terms of structural and functional measures [22]. Such inconsistency in the results may be attributed to the treatment duration as the regeneration of RGCs may need a longer time for observable neuroprotective effects. Nevertheless, based on the good safety profile demonstrated by clinical trials, we could still expect a potential neuroprotective effect if the treatment duration is designed to extend beyond 3 months.

Brain-derived neurotrophic factor enhances RGC survival by activating the extracellular signal-regulated kinases (Erk) Erk1/2 and c-jun, as well as inhibiting caspase 2 [23]. Recently, Cha et al. found that BDNF levels in serum and aqueous humor (AH) were significantly lower in patients with normal tension glaucoma (NTG) and primary open angle glaucoma (POAG) [24]. Clinical research conducted by Oddone et al. showed reduced serum levels of both BDNF and NGF in patients with early to moderate stages of glaucoma [20]. Uzel et al. also observed that BDNF in both serum and AH was lower in POAG patients, and the serum level of BDNF increased significantly three months after trabeculectomy [25]. These results suggest and reinforce the association between BDNF and glaucoma and that BDNF may serve as a potential biomarker for glaucoma detection and disease evaluation. In addition, several prior murine models also demonstrated that BDNF protects and facilitates RGCs' survival [26,27]. Most recently, Lazaldin et al. found that the intravitreal injection of BDNF can hinder RGC death caused by amyloid-β induced apoptosis in rats [28]. Nevertheless, more efforts are required to reach conclusions about the causal relationship between BDNF and glaucoma as well as whether the supplementation of BDNF is effective as a neuroprotective therapy for glaucoma.

The CNTF is expressed locally by RGCs. The concentration of CNTF is reduced in the AH and lacrimal fluid in patients with POAG [29]. NT-501, a polymeric device containing a genetically engineered human cell line that secretes CNTF and can be surgically implanted beneath the pars plana, has been trialed in retinal diseases without obvious treatment benefits [30,31]. Clinical trials of NT-501-encapsulated cell therapy are currently being undertaken to explore its therapeutic efficacy in the treatment of glaucoma (Clinical Trials ID NCT02862938 and NCT04577300).

While studies on NTFs show that they have great potential for neuroprotection, the challenge of clinical translation remains in how to accomplish effective and sustainable delivery to the retina. Intravitreal injection is a feasible method of delivering pure recombinant trophic factors to the retina, but in chronic diseases like glaucoma, which may last for several decades, this treatment modality may not be pragmatic if the injection has to be repeated frequently. Therefore, future studies may be directed toward the development of the sustained release of the intraocular implant containing NTFs or stem cell transplant that can modulate the levels of NTFs in the microenvironment. Alternatively, external therapies such as low-level electrical stimulation, in which the electrodes are attached around the eye and on the retina, can be used to induce local production of NTFs [32–34]. Previous rodent models have demonstrated the upregulation of CNTF and BDNF after electrical stimulation [35,36].

3. Ginkgo Biloba

Ginkgo is a traditional medication commonly used in Eastern countries. Modern research has been conducted to explore the neuroprotective effect of Ginkgo biloba extract (GBE) in the treatment of glaucoma, and several theories regarding the underlying mechanisms have been proposed. First, GBE may lead to increased blood flow by altering the blood viscosity and suppressing platelet-activating factors that can induce platelet aggregation, neutrophil degranulation, and ROS generation [37–39]. Second, the antioxidative capabilities of GBE may be exerted by its component poly-phenolic flavivonoids via the mechanism of reducing the oxidative stress in the mitochondria and scavenging free radicals [40–43]. Conflicting clinical results for the effectiveness of GBE have been reported in some clinical trials. Quaranta et al. reported improvements in preexisting visual field defects (average baseline mean deviation of 11.40 dB versus 8.78 dB after GBE treatment, $P = 0.0001$; average baseline-corrected pattern standard deviation of 10.93 dB versus 8.13 dB after treatment, $P = 0.0001$) after one month of 40 mg GBE oral capsule three times daily in patients with NTG without any topical hypotensive therapy [44]. Long-term treatment of 80 mg GBE twice daily also significantly slowed the progression of visual field defects, which improved from the pretreatment regression coefficient of mean deviation at −0.619 dB to −0.379 dB per year without affecting the IOP in NTG patients who concurrently used 1 or 2 hypotensive eyedrops [45]. However, a randomized controlled trial (RCT) revealed no effect of 40 mg GBE tablets three times daily for 4 weeks on visual field performance in NTG patients with insignificant IOP difference during the study period [46]. Although this RCT followed a similar study design to Quaranta's, the contradictory results may be attributed to some factors such as race and disease severity. In addition, the difference between Lee's and Guo's work cannot be compared directly due to different GBE dosages and treatment duration. Recently, Sabaner et al. demonstrated increased peripapillary vessel density on optical coherence tomography (OCT) angiography in healthy subjects after the four-week consumption of GBE 120mg oral capsule [47]. Because previous studies have demonstrated a positive correlation between peripapillary vessel density and visual field performance [48,49], a comprehensive research may be worth conducting to directly evaluate the changes in both vessel density and visual field in patients treated with GBE. Whether GBE is effective in the treatment of glaucoma may still need to be justified with further large clinical trials to identify patient characteristics, disease extent, concurrent treatments, etc. associated with the beneficial effects of GBE in patients with NTG.

4. Brimonidine

Aside from its well-known effect of decreasing IOP, brimonidine has shown neuroprotective properties against RGC death in several preclinical studies [50–53]. It has been shown to increase NTFs and alter N-methyl-D-aspartate (NMDA) receptors involved in glutamate toxicity [54,55]. It is reported to boost BDNF expression in RGCs, resulting in a neuroprotective effect [54]. In addition, brimonidine has been shown to exert neuroprotection by interfering with the amyloid-β pathway and lowering its levels since the disruption of amyloid precursor protein homeostasis and the subsequent accumulation of amyloid-β and its cytotoxicity may contribute to the death of the RGCs [56,57]. In clinical studies, brimonidine monotherapy has been shown to reduce the incidence of visual field progression compared with timolol in treated individuals (9% versus 30%) in the Low Pressure Glaucoma Study Group over 30 months, despite identical IOP-lowering effects [58]. However, the study was restricted by high dropout rates of 55% (54/99) in the brimonidine group and 29% (23/79) in the timolol group, as well as a relatively small sample size involved in the final analysis [59,60]. Topical brimonidine 0.2% administered over 3 months was also observed to increase contrast sensitivity; however, treatment with timolol 0.5% had no benefit, despite identical IOP reduction effects [61]. In addition, Tsai et al. reported no significant change in retinal nerve fiber layer (RNFL) thickness after the administration of brimonidine 0.2 % versus a statistically significant decrease in average RNFL thickness ($P = 0.004$) from baseline in the timolol 0.5 % group in patients with ocular hypertension treated for 1 year, despite similar mean IOP reduction in both groups [62]. Overall, these findings imply that brimonidine has a neuroprotective effect that is not related to IOP and may be used more widely for its IOP-independent treatment effect in glaucoma.

5. Calcium Channel Blocker (CCB)

Over the past few decades, calcium dysregulation has been regarded as a pathophysiological component in the degeneration of RGCs [63]. Theoretically, CCBs protect RGCs by preventing cell death caused by calcium influx and increasing local blood flow in ischemic tissues by inducing vasodilation [63,64]. A cell culture study by Yamada et al. demonstrated that CCBs, including iganidipine, nimodipine, and lomerizine, can facilitate RGC viability to sustain hypoxic damage by blocking calcium ion influx into RGCs [65]. Meanwhile, another in vitro study also showed that nilvadipine may be capable of inhibiting glutamate-induced RGC apoptosis by interfering with calcium influx [66]. An RCT revealed that in systemically healthy patients with NTG, nilvadipine in a dosage of 2 mg twice daily may preserve the optic nerve structure as assessed by direct ophthalmoscopy, improve optic nerve head blood flow, and slow visual field progression compared with the placebo group [67]. Another member of CCB, brovincamine, has also been shown to be beneficial in improving visual field results and retarding disease progression in patients with NTG [68,69]. Recently, Duan et al. reported an improvement in ocular hemodynamics as well as visual field defects by nimodipine combined with latanoprost in OAG patients [70]. An increase in superficial macular capillary vessel density was also found in patients with NTG after consuming sixty mg of nimodipine for three months [71]. Despite the aforementioned benefits of CCB revealed in these small-scale studies, one should keep in mind that the influence of vasodilation induced by CCB may not sufficiently explain the neuroprotective effect because vasodilation may inversely direct blood flow away from the ischemic tissues and exacerbate the condition [72]. In addition, reduced systemic blood pressure appears to decrease ONH blood flow and cause further damage to the optic nerve in patients with glaucoma [73]. As a result, future studies may be directed at evaluating the optimal dosage and improving the selectiveness of CCB to exert maximal neuroprotection while minimizing accompanying side effects.

6. Memantine

Glutamate is a neurotransmitter activating proapoptotic cascades via NMDA and non-NMDA receptors. Increased glutamate level has been thought to be a possible cause of glaucoma [74]. Memantine is an NMDA receptor antagonist that inhibits excessive glutamate activity, has been studied in experimental models of glaucoma, and has demonstrated a protective effect on RGC survival and reductions in functional loss [74–77]. Despite promising results from animal models, a large phase 3 RCT showed that daily treatment of memantine over 4 years has little impact on delaying visual field progression in patients with bilateral open-angle glaucoma [78]. Several variables influencing therapeutic effectiveness may be taken into account to explain the discrepancy between animal studies and clinical trials, including baseline glaucoma severity, trial duration, memantine dosage, and administration route. In addition, because glaucoma is a multifactorial disease, such complexity may be impossible to simulate by a single animal model. Additional factors that could affect the treatment outcome should also be evaluated, including the presence of disc hemorrhage, central corneal thickness, etc. Further research may be directed toward enrolling patients with early-stage glaucoma, a less heterogeneous population in terms of different glaucoma progression risk factors, longer study duration, and other doses and methods for medication delivery. Although the clinical study failed to prove the association between memantine and visual field outcome, studies investigating the changes in RNFL using OCT and OCT angiography are also warranted as they represent more objective structural parameters in the assessment and detection of early glaucoma [79,80].

7. Citicoline

Citicoline is a natural compound that participates in the chemical reaction involving the neurotransmitter acetylcholine and other neuronal membrane components. It is crucial for preserving the levels of sphingomyelin and cardiolipin in neurons [81]. Citicoline exerts its neuroprotective effect by reducing glutamate excitotoxicity, lowering oxidative stress in RGC damage, and improving axonal transport deficit [82,83]. While the axons of RGC in the retrobulbar space are rich in myelin, citicoline may be a potential treatment option to modulate RGC viability via phospholipid metabolism in the myelin membrane [84].

In an animal model of glaucoma using adult rats with optic nerve crush, a decrease in RGC density was attenuated following intraperitoneal citicoline administration, indicating a protective effect against neuronal degeneration [85]. Improvements in pattern electroretinogram (PERG) and visual evoked potentials (VEP) were demonstrated in an RCT enrolling participants with POAG following the intramuscular injection of citicoline 1000 mg/day for 60 days [86]. Similar findings were also observed in glaucoma patients treated with topical citicoline eyedrops [87,88]. A study consisting of 47 POAG patients with beta-blocker monotherapy revealed that patients treated with topical citicoline (OMK1®, Omikron Italia, 3 drops/day) over 4 months had improved PERG and VEP, whereas no significant changes in PERG and VEP were observed in the control group treated with beta-blocker monotherapy [87]. In addition, 4 cycles of treatment, in which each cycle contains oral citicoline solution 1 vial (500 mg of citicoline) for 4 months and stopped for 2 months, were effective in halting visual field progression in 41 patients with POAG despite well-controlled IOP with hypotensive medications [89]. Recently, an RCT revealed that the addition of citicoline eyedrops 3 times daily for 3 years in patients with progressing OAG might be effective in terms of reducing the progression of mean deviation on 10-2 visual field and RNFL thickness [90]. Based on the aforementioned findings, it appears that citicoline offers great potential as a future therapeutic strategy for glaucoma and other neurological illnesses. Several trials are currently underway to explore the treatment effect of citicoline and its related products. (Clinical Trials ID NCT05315206, NCT04499157, NCT04784234).

8. Antioxidant Q10

Coenzyme Q10 (CoQ10) is an important endogenous antioxidant and electron transport chain component [91]. The intraocular administration of CoQ10 therapies has been demonstrated to help delay RGC apoptosis and reduce glutamate concentration in a rat model [92]. A diet containing CoQ10 has also been shown to be neuroprotective against NMDA-induced retinal damage both in vitro and in mice in vivo [93]. In another mouse model of glaucoma, a diet supplemented with CoQ10 could reduce glutamate excitotoxicity and oxidative stress-mediated RGC degeneration and improve RGC survival by 29% [94]. A similar neuroprotective effect was also observed in a rat model given topical CoQ10 and vitamin E treatment for 4 weeks after mechanic optic nerve injury [95]. A small study conducted by Parisi et al. revealed that in patients with POAG, the administration of CoQ10 plus vitamin E had a beneficial effect on the inner retinal function and visual cortical responses by using PERG and VEP, respectively [96]. In an RCT containing 64 eyes with pseudoexfoliative glaucoma, a lower level of superoxide dismutase in the AH was found in patients treated with topical CoQun solution containing CoQ10 and vitamin E [97]. Currently, a large multicenter RCT evaluating the neuroprotective effects of CoQun eyedrops in addition to prostaglandin monotherapy on glaucoma progression in patients with POAG is still undergoing (Clinical Trials ID NCT03611530) [98]. As more studies are being conducted, we can expect the treatment potential of CoQ10 in conjunction with IOP-lowering medications in the management of glaucoma.

9. Nicotinamide (Vitamin B3)

Nicotinamide, also known as vitamin B3, is a precursor to nicotinamide adenine dinucleotide (NAD), which is a co-enzyme found in living cells and an essential molecule for the proper function of the metabolic system. As reduced NAD levels and mitochondrial dysfunctions are considered to be the hallmarks of the aging process [99], the question remains whether the repletion of nicotinamide could be beneficial in the treatment of neurodegenerative diseases including glaucoma. In 2017, Williams et al. first demonstrated that oral supplementation with high-dose nicotinamide could alleviate the decreased NAD levels in the retina caused by aging in DBA/2 J mice, a mouse model of age-dependent inherited glaucoma [100]. In line with the findings of NAD depletion in the murine model, Nzoughet et al. revealed a reduced plasma level of NAD in patients with POAG compared with the control group [100,101]. This finding further suggests the potential role of nicotinamide supplementation in treating human glaucoma. Therefore, to evaluate the effect of supplementation of nicotinamide in human glaucoma, Hui et al. examined 57 participants with early to moderate glaucoma with well-controlled IOP who received either oral placebo or nicotinamide at the dosage of 1.5 g/day for 6 weeks and 3 g/day for another 6 weeks, and all participants crossed over without washout after 12 weeks. Overall, they found that patients consuming nicotinamide had improved RGC function as evaluated by ERG, independent of IOP [102]. A recent phase 2 RCT also showed a greater number of improving visual field test locations and improved rates of change of pattern standard deviation in patients with treated, moderate open-angle glaucoma taking a combination of oral nicotinamide and pyruvate [103]. These findings emphasize the importance of nicotinamide supplementation in glaucoma treatment, and the ongoing clinical trials examining the long-term effects of high-dose nicotinamide in glaucoma will be crucial in establishing the therapeutic role of NAM supplementation to slow the loss of visual field in people with glaucoma. (Clinical Trials ID NCT05275738, NCT05405868).

10. Statins

Statins are medications originally used to treat hypercholesterolemia. Their major mechanism of action is to block HMG-CoA reductase and hence reduce cholesterol production. A recently published meta-analysis of observational studies reported that glaucoma was linked to high total cholesterol and low high-density lipoprotein levels [104], which strengthens the importance of blood lipid levels in the treatment of glaucoma. A rat model of chronic ocular hypertension was used to evaluate the neuroprotective effect of statins and showed the improved survival of RGCs, reduced apoptosis, and the suppression of glial activation in the retina in the statins group [105]. One study using confocal scanning laser ophthalmoscopy to measure the rate of progression of optic nerve parameters in glaucoma suspects taking statin showed a slowed progression of loss of rim volume, cross-sectional area of RNFL, and mean global RNFL thickness when compared with the control group [106]. Despite promising preclinical data, Kang et al. found no significant association between statin use and rates of change in mean deviation and RNFL thickness in those with glaucoma or glaucoma suspects [107]. The most recent meta-analysis revealed a slightly lower risk of OAG onset after using statins, whereas the association between statin use and OAG progression is still uncertain [108]. With the increasing application of statins in the treatment of cardiovascular and cerebrovascular diseases [109,110], more preclinical and clinical studies are warranted to elucidate the underlying neuroprotective mechanisms of statins against glaucomatous optic neuropathy.

11. Stem Cell Therapy

Stem cell therapy is gaining in popularity for its potential to treat neurodegenerative diseases such as glaucoma. In glaucoma, the ultimate goal is to restore vision by the neuroregeneration of injured or dead RGCs and their axons. Stem cell treatment may be therapeutic for glaucoma through two different mechanisms: (1) regenerating RGCs and producing new cells of different kinds. (2) providing a favorable neurotrophic environment to the damaged RGCs [111,112]. In addition, the RGCs are the ideal target for stem cell therapy because they have the benefit of being confined to the intraocular spaces and may be less likely to be affected by immune rejection [113].

Mesenchymal stem cells (MSCs) are multipotent and can differentiate into neurons and glial cells, support neuronal growth and synaptic connection, induce angiogenesis, modulate inflammatory responses, and reduce demyelination and apoptosis, which all contribute to their neuroprotective and regenerative effects [114]. The transplanted MSCs can secret various NTFs to promote cell survival, including CNTF (a potent RGC survival factor), bFGF (a simulator of axonal growth), GDNF, and BDNF [112,115]. Various IOP-dependent animal models of glaucoma have demonstrated effectiveness in terms of promoting RGC survival and reducing RGC loss via the intravitreal injection of MSCs [116–120] and protecting trabecular meshwork tissue via the intracameral injection of MSCs [121]. Recently, a clinical trial by Vivela et al. reported no significant improvement in visual performance or ERG in a patient with advanced glaucoma after the intravitreal injection of autologous bone marrow-derived MSCs. In addition, the development of retinal detachment with proliferative vitreoretinopathy was noted in another participant [122]. Therefore, despite successful outcomes shown in animal models, there are still obstacles that can hinder the clinical translation of stem cell therapy into human application. Particularly, the complexity of human disease states may not be exactly represented by a controlled experimental environment in animal models. Nevertheless, larger clinical trials enrolling more participants with different disease severity; using different administration routes, i.e., intracamerally or intravitreally; and following for a longer period of duration are still warranted to fully elucidate the clinical effectiveness of this treatment modality.

Some safety issues need to be addressed before stem cell therapy could be successfully used in clinical practice. First, the balance between graft survival and tumorigenesis must be carefully assessed because the longer the stem cell lasts, the more likely the tumor might develop. Thus, meticulous laboratory and clinical research will be required to guarantee that the potential benefit of neuroprotection far outperforms the possible danger of tumor induction. Second, the implanted cells not only release ideal and desired trophic factors for supporting RGCs, but they may also secret other agents that could be potentially detrimental to the microenvironment of RGC [123]. Third, the differences in efficacy among various animal models for glaucomatous optic neuropathy may need to be justified before clinical translation. Therefore, more efforts are required before stem cell therapy can become practical in clinical settings.

12. Gene Therapy

Gene therapy has made remarkable progress in the past few decades. It offers the potential to help patients with damaged RGCs regain their lost vision. Clinical trials conducted on patients with congenital retinal diseases such as Leber hereditary optic neuropathy have shown promising results with direct gene medical application in the treatment of optic neuropathy [124,125]. However, genetic treatment in the context of glaucoma remains challenging due to its multifactorial and polygenic properties.

Some animal studies have demonstrated the efficacy of gene therapy in the treatment of glaucoma. For example, experimental studies conducted by Jain et al. showed that clustered regularly interspaced short palindromic repeats (CRISPR)-mediated genome editing of myocilin (MYOC)-dominant gain-of-function mutations effectively lowered IOP and hindered glaucomatous damage by inducing the loss of function of mutant MYOC in a mouse model of MYOC-associated POAG [126]. Another promising result was shown in a recent study using multiple rodent glaucoma models, in which the reactivation of CaMKII activity via the intravitreal injection of adeno-associated virus (AAV) vectors in diseased mice protects RGCs and preserves visual function and visually guided behavior [114]. Gene therapy also exerts its neuroprotective effect through encoding NTFs such as BDNF and CNTF [127,128]. In a rat model of optic nerve injury, Osborne et al. demonstrated enhancements of RGC survival and no significant adverse effects on the retinal structure or electrophysiological performance following the intravitreal injection of AAV2 TrkB-2A-mBDNF [127]. Various gene targets have also been studied in the experimental models of glaucoma, including BCLXL, NMNAT2, Myc-associated protein X, and XIAP [129–132].

With the advancement of whole-genome sequencing and genome editing technology, further genes related to the pathogenesis of glaucoma will be able to be discovered and tested as potential therapeutic targets. While there are still many obstacles to overcome before glaucoma gene therapy becomes clinically available, the progress in understanding the genetic etiology of glaucoma and breakthroughs in RGC neuroprotection in various animal models still hold the potential for leading to a new frontier of gene therapy in glaucoma.

13. Conclusions

Neuroprotection has the potential to play a critical role in glaucoma treatment. Improvement in RGCs survival and a decrease in cell death can not only slow disease progression but even restore visual function through tissue regeneration. Although several treatment modalities exhibit neuroprotective effects in experimental or clinical studies regarding glaucoma, only a few of them have resulted in approved therapy clinically, and the road to glaucoma neuroprotection remains long. Supplementary Table S1 summarizes the essence of the clinical study of each treatment modality and its implications. Overall, in clinical scenarios such as patients who are intolerant to hypotensive medications or unwilling to receive laser or surgical intervention or who have progressive glaucomatous defect despite well-controlled IOP, supplements such as GBE, citicoline, antioxidant Q10, and nicotinamide may be considered if they are available and not harmful to the patient's health. Brimonidine may also be used at the physician's discretion. The clinical usage and effectiveness of CCB, memantine, and statin remain to be justified. Neurotrophic factors, stem cell therapy, and gene therapy warrant further investigation before they can be administered to patients. Advancement in the evolution of neuroprotective therapy will be aided by substantial investment in genetic and biomolecular research.

Supplementary Materials: The following supporting information can be downloaded at: https://www.mdpi.com/article/10.3390/jpm12111884/s1, Table S1: Summary table of the included clinical studies.

Author Contributions: Conceptualization, C.-Y.K. and C.J.-L.L.; writing—original draft preparation, C.-Y.K.; writing—review and editing, C.J.-L.L.; supervision, C.J.-L.L. All authors have read and agreed to the published version of the manuscript.

Funding: This research received no external funding.

Conflicts of Interest: The authors declare no conflict of interest.

References

1. Weinreb, R.N.; Aung, T.; Medeiros, F.A. The pathophysiology and treatment of glaucoma: A review. *JAMA* **2014**, *311*, 1901–1911. [CrossRef] [PubMed]
2. Garway-Heath, D.F.; Lascaratos, G.; Bunce, C.; Crabb, D.P.; Russell, R.A.; Shah, A.; United Kingdom Glaucoma Treatment Study Investigators. The United Kingdom Glaucoma Treatment Study: A multicenter, randomized, placebo-controlled clinical trial: Design and methodology. *Ophthalmology* **2013**, *120*, 68–76. [CrossRef] [PubMed]
3. Heijl, A.; Leske, M.C.; Bengtsson, B.; Hyman, L.; Bengtsson, B.; Hussein, M. Reduction of intraocular pressure and glaucoma progression: Results from the Early Manifest Glaucoma Trial. *Arch. Ophthalmol.* **2002**, *120*, 1268–1279. [CrossRef] [PubMed]
4. Kass, M.A.; Heuer, D.K.; Higginbotham, E.J.; Johnson, C.A.; Keltner, J.L.; Miller, J.P.; Parrish, R.K., 2nd; Wilson, M.R.; Gordon, M.O. The Ocular Hypertension Treatment Study: A randomized trial determines that topical ocular hypotensive medication delays or prevents the onset of primary open-angle glaucoma. *Arch. Ophthalmol.* **2002**, *120*, 701–713; Discussion 730–829. [CrossRef]

5. Collaborative Normal-Tension Glaucoma Study Group. Comparison of glaucomatous progression between untreated patients with normal-tension glaucoma and patients with therapeutically reduced intraocular pressures. *Am. J. Ophthalmol.* **1998**, *126*, 487–497. [CrossRef]
6. Chidlow, G.; Ebneter, A.; Wood, J.P.; Casson, R.J. The optic nerve head is the site of axonal transport disruption, axonal cytoskeleton damage and putative axonal regeneration failure in a rat model of glaucoma. *Acta Neuropathol.* **2011**, *121*, 737–751. [CrossRef]
7. Quigley, H.A.; Addicks, E.M.; Green, W.R.; Maumenee, A.E. Optic nerve damage in human glaucoma. II. The site of injury and susceptibility to damage. *Arch. Ophthalmol.* **1981**, *99*, 635–649. [CrossRef]
8. Casson, R.J.; Chidlow, G.; Ebneter, A.; Wood, J.P.; Crowston, J.; Goldberg, I. Translational neuroprotection research in glaucoma: A review of definitions and principles. *Clin. Exp. Ophthalmol.* **2012**, *40*, 350–357. [CrossRef]
9. Weinreb, R.N.; Levin, L.A. Is neuroprotection a viable therapy for glaucoma? *Arch. Ophthalmol.* **1999**, *117*, 1540–1544. [CrossRef]
10. Pietrucha-Dutczak, M.; Amadio, M.; Govoni, S.; Lewin-Kowalik, J.; Smedowski, A. The Role of Endogenous Neuroprotective Mechanisms in the Prevention of Retinal Ganglion Cells Degeneration. *Front. Neurosci.* **2018**, *12*, 834. [CrossRef]
11. Barbacid, M. Neurotrophic factors and their receptors. *Curr. Opin. Cell Biol.* **1995**, *7*, 148–155. [CrossRef]
12. Kimura, A.; Namekata, K.; Guo, X.; Harada, C.; Harada, T. Neuroprotection, Growth Factors and BDNF-TrkB Signalling in Retinal Degeneration. *Int. J. Mol. Sci.* **2016**, *17*, 1584. [CrossRef] [PubMed]
13. Chang, E.E.; Goldberg, J.L. Glaucoma 2.0: Neuroprotection, neuroregeneration, neuroenhancement. *Ophthalmology* **2012**, *119*, 979–986. [CrossRef] [PubMed]
14. Mirowska-Guzel, D. The role of neurotrophic factors in the pathology and treatment of multiple sclerosis. *Immunopharmacol. Immunotoxicol.* **2009**, *31*, 32–38. [CrossRef]
15. Nasrolahi, A.; Javaherforooshzadeh, F.; Jafarzadeh-Gharehziaaddin, M.; Mahmoudi, J.; Asl, K.D.; Shabani, Z. Therapeutic potential of neurotrophic factors in Alzheimer's Disease. *Mol. Biol. Rep.* **2022**, *49*, 2345–2357. [CrossRef]
16. Sullivan, A.M.; Toulouse, A. Neurotrophic factors for the treatment of Parkinson's disease. *Cytokine Growth Factor Rev.* **2011**, *22*, 157–165. [CrossRef]
17. Ji, J.Z.; Elyaman, W.; Yip, H.K.; Lee, V.W.; Yick, L.W.; Hugon, J.; So, K.F. CNTF promotes survival of retinal ganglion cells after induction of ocular hypertension in rats: The possible involvement of STAT3 pathway. *Eur. J. Neurosci.* **2004**, *19*, 265–272. [CrossRef]
18. Koeberle, P.D.; Ball, A.K. Neurturin enhances the survival of axotomized retinal ganglion cells in vivo: Combined effects with glial cell line-derived neurotrophic factor and brain-derived neurotrophic factor. *Neuroscience* **2002**, *110*, 555–567. [CrossRef]
19. Lambiase, A.; Aloe, L.; Centofanti, M.; Parisi, V.; Báo, S.N.; Mantelli, F.; Colafrancesco, V.; Manni, G.L.; Bucci, M.G.; Bonini, S.; et al. Experimental and clinical evidence of neuroprotection by nerve growth factor eye drops: Implications for glaucoma. *Proc. Natl. Acad. Sci. USA* **2009**, *106*, 13469–13474. [CrossRef]
20. Oddone, F.; Roberti, G.; Micera, A.; Busanello, A.; Bonini, S.; Quaranta, L.; Agnifili, L.; Manni, G. Exploring Serum Levels of Brain Derived Neurotrophic Factor and Nerve Growth Factor Across Glaucoma Stages. *PLoS ONE* **2017**, *12*, e0168565. [CrossRef]
21. Schuettauf, F.; Vorwerk, C.; Naskar, R.; Orlin, A.; Quinto, K.; Zurakowski, D.; Dejneka, N.S.; Klein, R.L.; Meyer, E.M.; Bennett, J. Adeno-associated viruses containing bFGF or BDNF are neuroprotective against excitotoxicity. *Curr. Eye Res.* **2004**, *29*, 379–386. [CrossRef]
22. Beykin, G.; Stell, L.; Halim, M.S.; Nuñez, M.; Popova, L.; Nguyen, B.T.; Groth, S.L.; Dennis, A.; Li, Z.; Atkins, M.; et al. Phase 1b Randomized Controlled Study of Short Course Topical Recombinant Human Nerve Growth Factor (rhNGF) for Neuroenhancement in Glaucoma: Safety, Tolerability, and Efficacy Measure Outcomes. *Am. J. Ophthalmol.* **2022**, *234*, 223–234. [CrossRef]
23. Johnson, T.V.; Bull, N.D.; Martin, K.R. Neurotrophic factor delivery as a protective treatment for glaucoma. *Exp. Eye Res.* **2011**, *93*, 196–203. [CrossRef]
24. Cha, Y.W.; Kim, S.T. Serum and aqueous humor levels of brain-derived neurotrophic factor in patients with primary open-angle glaucoma and normal-tension glaucoma. *Int. Ophthalmol.* **2021**, *41*, 3869–3875. [CrossRef]
25. Uzel, M.M.; Elgin, U.; Boral, B.; Çiçek, M.; Şen, E.; Şener, B.; Yılmazbaş, P. The effect of trabeculectomy on serum brain-derived neurotrophic factor levels in primary open-angle glaucoma. *Graefe's Arch. Clin. Exp. Ophthalmol.* **2018**, *256*, 1173–1178. [CrossRef]
26. Ko, M.L.; Hu, D.N.; Ritch, R.; Sharma, S.C.; Chen, C.F. Patterns of retinal ganglion cell survival after brain-derived neurotrophic factor administration in hypertensive eyes of rats. *Neurosci. Lett.* **2001**, *305*, 139–142. [CrossRef]
27. Peinado-Ramón, P.; Salvador, M.; Villegas-Pérez, M.P.; Vidal-Sanz, M. Effects of axotomy and intraocular administration of NT-4, NT-3, and brain-derived neurotrophic factor on the survival of adult rat retinal ganglion cells. A quantitative in vivo study. *Investig. Ophthalmol. Vis. Sci.* **1996**, *37*, 489–500.
28. Lazaldin, M.A.M.; Iezhitsa, I.; Agarwal, R.; Agarwal, P.; Ismail, N.M. Neuroprotective effects of exogenous brain-derived neurotrophic factor on amyloid-beta 1-40-induced retinal degeneration. *Neural Regen. Res.* **2023**, *18*, 382–388. [CrossRef]
29. Shpak, A.A.; Guekht, A.B.; Druzhkova, T.A.; Kozlova, K.I.; Gulyaeva, N.V. Ciliary neurotrophic factor in patients with primary open-angle glaucoma and age-related cataract. *Mol. Vis.* **2017**, *23*, 799–809.
30. Birch, D.G.; Bennett, L.D.; Duncan, J.L.; Weleber, R.G.; Pennesi, M.E. Long-term Follow-up of Patients With Retinitis Pigmentosa Receiving Intraocular Ciliary Neurotrophic Factor Implants. *Am. J. Ophthalmol.* **2016**, *170*, 10–14. [CrossRef]

31. Thanos, C.G.; Bell, W.J.; O'Rourke, P.; Kauper, K.; Sherman, S.; Stabila, P.; Tao, W. Sustained secretion of ciliary neurotrophic factor to the vitreous, using the encapsulated cell therapy-based NT-501 intraocular device. *Tissue Eng.* **2004**, *10*, 1617–1622. [CrossRef]
32. Gordon, T. Electrical Stimulation to Enhance Axon Regeneration After Peripheral Nerve Injuries in Animal Models and Humans. *Neurotherapeutics* **2016**, *13*, 295–310. [CrossRef]
33. Pardue, M.T.; Allen, R.S. Neuroprotective strategies for retinal disease. *Prog. Retin. Eye Res.* **2018**, *65*, 50–76. [CrossRef]
34. Sehic, A.; Guo, S.; Cho, K.S.; Corraya, R.M.; Chen, D.F.; Utheim, T.P. Electrical Stimulation as a Means for Improving Vision. *Am. J. Pathol.* **2016**, *186*, 2783–2797. [CrossRef]
35. Hanif, A.M.; Kim, M.K.; Thomas, J.G.; Ciavatta, V.T.; Chrenek, M.; Hetling, J.R.; Pardue, M.T. Whole-eye electrical stimulation therapy preserves visual function and structure in P23H-1 rats. *Exp. Eye Res.* **2016**, *149*, 75–83. [CrossRef]
36. Ni, Y.Q.; Gan, D.K.; Xu, H.D.; Xu, G.Z.; Da, C.D. Neuroprotective effect of transcorneal electrical stimulation on light-induced photoreceptor degeneration. *Exp. Neurol.* **2009**, *219*, 439–452. [CrossRef] [PubMed]
37. Akiba, S.; Kawauchi, T.; Oka, T.; Hashizume, T.; Sato, T. Inhibitory effect of the leaf extract of Ginkgo biloba L. on oxidative stress-induced platelet aggregation. *Biochem. Mol. Biol. Int.* **1998**, *46*, 1243–1248. [CrossRef]
38. Chung, H.S.; Harris, A.; Kristinsson, J.K.; Ciulla, T.A.; Kagemann, C.; Ritch, R. Ginkgo biloba extract increases ocular blood flow velocity. *J. Ocul. Pharmacol. Ther.* **1999**, *15*, 233–240. [CrossRef]
39. Kleijnen, J.; Knipschild, P. Ginkgo biloba. *Lancet* **1992**, *340*, 1136–1139. [CrossRef]
40. Cheung, Z.H.; So, K.F.; Lu, Q.; Yip, H.K.; Wu, W.; Shan, J.J.; Pang, P.K.; Chen, C.F. Enhanced survival and regeneration of axotomized retinal ganglion cells by a mixture of herbal extracts. *J. Neurotrauma* **2002**, *19*, 369–378. [CrossRef] [PubMed]
41. Marcocci, L.; Maguire, J.J.; Droy-Lefaix, M.T.; Packer, L. The nitric oxide-scavenging properties of Ginkgo biloba extract EGb 761. *Biochem. Biophys. Res. Commun.* **1994**, *201*, 748–755. [CrossRef] [PubMed]
42. Ritch, R. Potential role for Ginkgo biloba extract in the treatment of glaucoma. *Med. Hypotheses* **2000**, *54*, 221–235. [CrossRef] [PubMed]
43. Sacca, S.C.; Pascotto, A.; Camicione, P.; Capris, P.; Izzotti, A. Oxidative DNA damage in the human trabecular meshwork: Clinical correlation in patients with primary open-angle glaucoma. *Arch. Ophthalmol.* **2005**, *123*, 458–463. [CrossRef] [PubMed]
44. Quaranta, L.; Bettelli, S.; Uva, M.G.; Semeraro, F.; Turano, R.; Gandolfo, E. Effect of Ginkgo biloba extract on preexisting visual field damage in normal tension glaucoma. *Ophthalmology* **2003**, *110*, 359–362; discussion 362–364. [CrossRef]
45. Lee, J.; Sohn, S.W.; Kee, C. Effect of Ginkgo biloba extract on visual field progression in normal tension glaucoma. *J. Glaucoma* **2013**, *22*, 780–784. [CrossRef]
46. Guo, X.; Kong, X.; Huang, R.; Jin, L.; Ding, X.; He, M.; Liu, X.; Patel, M.C.; Congdon, N.G. Effect of Ginkgo biloba on visual field and contrast sensitivity in Chinese patients with normal tension glaucoma: A randomized, crossover clinical trial. *Investig. Ophthalmol. Vis. Sci.* **2014**, *55*, 110–116. [CrossRef]
47. Sabaner, M.C.; Dogan, M.; Altin, S.S.; Balaman, C.; Yilmaz, C.; Omur, A.; Zeybek, I.; Palaz, M. Ginkgo Biloba affects microvascular morphology: A prospective optical coherence tomography angiography pilot study. *Int. Ophthalmol.* **2021**, *41*, 1053–1061. [CrossRef]
48. Lin, Y.H.; Huang, S.M.; Yeung, L.; Ku, W.C.; Chen, H.S.; Lai, C.C.; Chuang, L.H. Correlation of Visual Field With Peripapillary Vessel Density Through Optical Coherence Tomography Angiography in Normal-Tension Glaucoma. *Trans. Vis. Sci. Technol.* **2020**, *9*, 26. [CrossRef]
49. Zhong, H.; Dong, Q.; Cun, Q.; He, G.; Tao, Y.; Song, K.; Lu, Y.; Zhu, Q.; Chen, X.; Chen, Q. Peripapillary vessel density correlates with visual field mean sensitivity in highly myopic eyes. *J. Transl. Med.* **2022**, *20*, 119. [CrossRef]
50. Hernández, M.; Urcola, J.H.; Vecino, E. Retinal ganglion cell neuroprotection in a rat model of glaucoma following brimonidine, latanoprost or combined treatments. *Exp. Eye Res.* **2008**, *86*, 798–806. [CrossRef]
51. Pinar-Sueiro, S.; Urcola, H.; Rivas, M.A.; Vecino, E. Prevention of retinal ganglion cell swelling by systemic brimonidine in a rat experimental glaucoma model. *Clin. Exp. Ophthalmol.* **2011**, *39*, 799–807. [CrossRef] [PubMed]
52. Yoles, E.; Wheeler, L.A.; Schwartz, M. Alpha2-adrenoreceptor agonists are neuroprotective in a rat model of optic nerve degeneration. *Investig. Ophthalmol. Vis. Sci.* **1999**, *40*, 65–73.
53. Conti, F.; Romano, G.L.; Eandi, C.M.; Toro, M.D.; Rejdak, R.; Di Benedetto, G.; Lazzara, F.; Bernardini, R.; Drago, F.; Cantarella, G.; et al. Brimonidine is Neuroprotective in Animal Paradigm of Retinal Ganglion Cell Damage. *Front. Pharmacol.* **2021**, *12*, 705405. [CrossRef]
54. Gao, H.; Qiao, X.; Cantor, L.B.; WuDunn, D. Up-regulation of brain-derived neurotrophic factor expression by brimonidine in rat retinal ganglion cells. *Arch. Ophthalmol.* **2002**, *120*, 797–803. [CrossRef] [PubMed]
55. Jung, K.I.; Kim, J.H.; Park, C.K. α2-Adrenergic modulation of the glutamate receptor and transporter function in a chronic ocular hypertension model. *Eur. J. Pharmacol.* **2015**, *765*, 274–283. [CrossRef]
56. Goldblum, D.; Kipfer-Kauer, A.; Sarra, G.M.; Wolf, S.; Frueh, B.E. Distribution of amyloid precursor protein and amyloid-beta immunoreactivity in DBA/2J glaucomatous mouse retinas. *Investig. Ophthalmol. Vis. Sci.* **2007**, *48*, 5085–5090. [CrossRef]
57. Nizari, S.; Guo, L.; Davis, B.M.; Normando, E.M.; Galvao, J.; Turner, L.A.; Bizrah, M.; Dehabadi, M.; Tian, K.; Cordeiro, M.F. Non-amyloidogenic effects of α2 adrenergic agonists: Implications for brimonidine-mediated neuroprotection. *Cell Death Dis.* **2016**, *7*, e2514. [CrossRef]
58. Krupin, T.; Liebmann, J.M.; Greenfield, D.S.; Ritch, R.; Gardiner, S. A randomized trial of brimonidine versus timolol in preserving visual function: Results from the Low-Pressure Glaucoma Treatment Study. *Am. J. Ophthalmol.* **2011**, *151*, 671–681. [CrossRef]

59. Garudadri, C.S.; Choudhari, N.S.; Rao, H.L.; Senthil, S. A randomized trial of brimonidine versus timolol in preserving visual function: Results from the Low-pressure Glaucoma Treatment Study. *Am. J. Ophthalmol.* **2011**, *152*, 877; Author Reply 877–878. [CrossRef]
60. Sena, D.F.; Lindsley, K. Neuroprotection for treatment of glaucoma in adults. *Cochrane Database Syst. Rev.* **2017**, *1*, CD006539. [CrossRef]
61. Evans, D.W.; Hosking, S.L.; Gherghel, D.; Bartlett, J.D. Contrast sensitivity improves after brimonidine therapy in primary open angle glaucoma: A case for neuroprotection. *Br. J. Ophthalmol.* **2003**, *87*, 1463–1465. [CrossRef] [PubMed]
62. Tsai, J.C.; Chang, H.W. Comparison of the effects of brimonidine 0.2% and timolol 0.5% on retinal nerve fiber layer thickness in ocular hypertensive patients: A prospective, unmasked study. *J. Ocul. Pharmacol. Ther.* **2005**, *21*, 475–482. [CrossRef] [PubMed]
63. Crish, S.D.; Calkins, D.J. Neurodegeneration in glaucoma: Progression and calcium-dependent intracellular mechanisms. *Neuroscience* **2011**, *176*, 1–11. [CrossRef] [PubMed]
64. Osborne, N.N.; Wood, J.P.; Chidlow, G.; Casson, R.; DeSantis, L.; Schmidt, K.G. Effectiveness of levobetaxolol and timolol at blunting retinal ischaemia is related to their calcium and sodium blocking activities: Relevance to glaucoma. *Brain Res. Bull.* **2004**, *62*, 525–528. [CrossRef]
65. Yamada, H.; Chen, Y.N.; Aihara, M.; Araie, M. Neuroprotective effect of calcium channel blocker against retinal ganglion cell damage under hypoxia. *Brain Res.* **2006**, *1071*, 75–80. [CrossRef] [PubMed]
66. Otori, Y.; Kusaka, S.; Kawasaki, A.; Morimura, H.; Miki, A.; Tano, Y. Protective effect of nilvadipine against glutamate neurotoxicity in purified retinal ganglion cells. *Brain Res.* **2003**, *961*, 213–219. [CrossRef]
67. Koseki, N.; Araie, M.; Tomidokoro, A.; Nagahara, M.; Hasegawa, T.; Tamaki, Y.; Yamamoto, S. A placebo-controlled 3-year study of a calcium blocker on visual field and ocular circulation in glaucoma with low-normal pressure. *Ophthalmology* **2008**, *115*, 2049–2057. [CrossRef]
68. Koseki, N.; Araie, M.; Yamagami, J.; Shirato, S.; Yamamoto, S. Effects of oral brovincamine on visual field damage in patients with normal-tension glaucoma with low-normal intraocular pressure. *J. Glaucoma* **1999**, *8*, 117–123. [CrossRef]
69. Sawada, A.; Kitazawa, Y.; Yamamoto, T.; Okabe, I.; Ichien, K. Prevention of visual field defect progression with brovincamine in eyes with normal-tension glaucoma. *Ophthalmology* **1996**, *103*, 283–288. [CrossRef]
70. Duan, H.P.; Liu, R. Efficacy of Nimodipine Combined with Latanoprost in Treating Open-Angle Glaucoma and Its Influence on Ocular Hemodynamics and Visual Field Defects. *Drug Des. Dev. Ther.* **2022**, *16*, 749–757. [CrossRef]
71. Hu, X.; Wang, X.; Dai, Y.; Qiu, C.; Shang, K.; Sun, X. Effect of Nimodipine on Macular and Peripapillary Capillary Vessel Density in Patients with Normal-tension Glaucoma Using Optical Coherence Tomography Angiography. *Curr. Eye Res.* **2021**, *46*, 1861–1866. [CrossRef] [PubMed]
72. Araie, M.; Mayama, C. Use of calcium channel blockers for glaucoma. *Prog. Retin. Eye Res.* **2011**, *30*, 54–71. [CrossRef] [PubMed]
73. Yilmaz, K.C.; Sur Gungor, S.; Ciftci, O.; Akman, A.; Muderrisoglu, H. Relationship between primary open angle glaucoma and blood pressure. *Acta Cardiol.* **2020**, *75*, 54–58. [CrossRef] [PubMed]
74. Hare, W.A.; WoldeMussie, E.; Lai, R.K.; Ton, H.; Ruiz, G.; Chun, T.; Wheeler, L. Efficacy and safety of memantine treatment for reduction of changes associated with experimental glaucoma in monkey, I: Functional measures. *Investig. Ophthalmol. Vis. Sci.* **2004**, *45*, 2625–2639. [CrossRef]
75. Hare, W.; WoldeMussie, E.; Lai, R.; Ton, H.; Ruiz, G.; Feldmann, B.; Wijono, M.; Chun, T.; Wheeler, L. Efficacy and safety of memantine, an NMDA-type open-channel blocker, for reduction of retinal injury associated with experimental glaucoma in rat and monkey. *Surv. Ophthalmol.* **2001**, *45* (Suppl. 3), S284–S289; Discussion S286–S295. [CrossRef]
76. WoldeMussie, E.; Yoles, E.; Schwartz, M.; Ruiz, G.; Wheeler, L.A. Neuroprotective effect of memantine in different retinal injury models in rats. *J. Glaucoma* **2002**, *11*, 474–480. [CrossRef]
77. Yücel, Y.H.; Gupta, N.; Zhang, Q.; Mizisin, A.P.; Kalichman, M.W.; Weinreb, R.N. Memantine protects neurons from shrinkage in the lateral geniculate nucleus in experimental glaucoma. *Arch. Ophthalmol.* **2006**, *124*, 217–225. [CrossRef]
78. Weinreb, R.N.; Liebmann, J.M.; Cioffi, G.A.; Goldberg, I.; Brandt, J.D.; Johnson, C.A.; Zangwill, L.M.; Schneider, S.; Badger, H.; Bejanian, M. Oral Memantine for the Treatment of Glaucoma: Design and Results of 2 Randomized, Placebo-Controlled, Phase 3 Studies. *Ophthalmology* **2018**, *125*, 1874–1885. [CrossRef]
79. Kromer, R.; Glusa, P.; Framme, C.; Pielen, A.; Junker, B. Optical coherence tomography angiography analysis of macular flow density in glaucoma. *Acta Ophthalmol.* **2019**, *97*, e199–e206. [CrossRef]
80. Nouri-Mahdavi, K.; Nikkhou, K.; Hoffman, D.C.; Law, S.K.; Caprioli, J. Detection of early glaucoma with optical coherence tomography (StratusOCT). *J. Glaucoma* **2008**, *17*, 183–188. [CrossRef]
81. Adibhatla, R.M.; Hatcher, J.F.; Dempsey, R.J. Effects of citicoline on phospholipid and glutathione levels in transient cerebral ischemia. *Stroke* **2001**, *32*, 2376–2381. [CrossRef] [PubMed]
82. Grieb, P.; Jünemann, A.; Rekas, M.; Rejdak, R. Citicoline: A Food Beneficial for Patients Suffering from or Threated with Glaucoma. *Front. Aging Neurosci.* **2016**, *8*, 73. [CrossRef] [PubMed]
83. Mir, C.; Clotet, J.; Aledo, R.; Durany, N.; Argemí, J.; Lozano, R.; Cervós-Navarro, J.; Casals, N. CDP-choline prevents glutamate-mediated cell death in cerebellar granule neurons. *J. Mol. Neurosci.* **2003**, *20*, 53–60. [CrossRef]
84. FitzGibbon, T.; Nestorovski, Z. Human intraretinal myelination: Axon diameters and axon/myelin thickness ratios. *Indian J. Ophthalmol.* **2013**, *61*, 567–575. [CrossRef] [PubMed]

85. Schuettauf, F.; Rejdak, R.; Thaler, S.; Bolz, S.; Lehaci, C.; Mankowska, A.; Zarnowski, T.; Junemann, A.; Zagorski, Z.; Zrenner, E.; et al. Citicoline and lithium rescue retinal ganglion cells following partial optic nerve crush in the rat. *Exp. Eye Res.* **2006**, *83*, 1128–1134. [CrossRef]
86. Parisi, V.; Manni, G.; Colacino, G.; Bucci, M.G. Cytidine-5′-diphosphocholine (citicoline) improves retinal and cortical responses in patients with glaucoma. *Ophthalmology* **1999**, *106*, 1126–1134. [CrossRef]
87. Parisi, V.; Centofanti, M.; Ziccardi, L.; Tanga, L.; Michelessi, M.; Roberti, G.; Manni, G. Treatment with citicoline eye drops enhances retinal function and neural conduction along the visual pathways in open angle glaucoma. *Graefe's Arch. Clin. Exp. Ophthalmol.* **2015**, *253*, 1327–1340. [CrossRef]
88. Roberti, G.; Tanga, L.; Parisi, V.; Sampalmieri, M.; Centofanti, M.; Manni, G. A preliminary study of the neuroprotective role of citicoline eye drops in glaucomatous optic neuropathy. *Indian J. Ophthalmol.* **2014**, *62*, 549–553. [CrossRef]
89. Ottobelli, L.; Manni, G.L.; Centofanti, M.; Iester, M.; Allevena, F.; Rossetti, L. Citicoline oral solution in glaucoma: Is there a role in slowing disease progression? *Ophthalmologica* **2013**, *229*, 219–226. [CrossRef]
90. Rossetti, L.; Iester, M.; Tranchina, L.; Ottobelli, L.; Coco, G.; Calcatelli, E.; Ancona, C.; Cirafici, P.; Manni, G. Can Treatment With Citicoline Eyedrops Reduce Progression in Glaucoma? The Results of a Randomized Placebo-controlled Clinical Trial. *J. Glaucoma* **2020**, *29*, 513–520. [CrossRef]
91. Edwards, G.; Lee, Y.; Kim, M.; Bhanvadia, S.; Kim, K.Y.; Ju, W.K. Effect of Ubiquinol on Glaucomatous Neurodegeneration and Oxidative Stress: Studies for Retinal Ganglion Cell Survival and/or Visual Function. *Antioxidants* **2020**, *9*, 952. [CrossRef] [PubMed]
92. Nucci, C.; Tartaglione, R.; Cerulli, A.; Mancino, R.; Spanò, A.; Cavaliere, F.; Rombolà, L.; Bagetta, G.; Corasaniti, M.T.; Morrone, L.A. Retinal damage caused by high intraocular pressure-induced transient ischemia is prevented by coenzyme Q10 in rat. In *International Review of Neurobiology*; Academic Press: Cambridge, MA, USA, 2007; Volume 82, pp. 397–406. [CrossRef]
93. Nakajima, Y.; Inokuchi, Y.; Nishi, M.; Shimazawa, M.; Otsubo, K.; Hara, H. Coenzyme Q10 protects retinal cells against oxidative stress in vitro and in vivo. *Brain Res.* **2008**, *1226*, 226–233. [CrossRef] [PubMed]
94. Lee, D.; Shim, M.S.; Kim, K.Y.; Noh, Y.H.; Kim, H.; Kim, S.Y.; Weinreb, R.N.; Ju, W.K. Coenzyme Q10 inhibits glutamate excitotoxicity and oxidative stress-mediated mitochondrial alteration in a mouse model of glaucoma. *Investig. Ophthalmol. Vis. Sci.* **2014**, *55*, 993–1005. [CrossRef] [PubMed]
95. Ekicier Acar, S.; Sarıcaoğlu, M.S.; Çolak, A.; Aktaş, Z.; Sepici Dinçel, A. Neuroprotective effects of topical coenzyme Q10 + vitamin E in mechanic optic nerve injury model. *Eur. J. Ophthalmol.* **2020**, *30*, 714–722. [CrossRef] [PubMed]
96. Parisi, V.; Centofanti, M.; Gandolfi, S.; Marangoni, D.; Rossetti, L.; Tanga, L.; Tardini, M.; Traina, S.; Ungaro, N.; Vetrugno, M.; et al. Effects of coenzyme Q10 in conjunction with vitamin E on retinal-evoked and cortical-evoked responses in patients with open-angle glaucoma. *J. Glaucoma* **2014**, *23*, 391–404. [CrossRef] [PubMed]
97. Ozates, S.; Elgin, K.U.; Yilmaz, N.S.; Demirel, O.O.; Sen, E.; Yilmazbas, P. Evaluation of oxidative stress in pseudo-exfoliative glaucoma patients treated with and without topical coenzyme Q10 and vitamin E. *Eur. J. Ophthalmol.* **2019**, *29*, 196–201. [CrossRef]
98. Quaranta, L.; Riva, I.; Biagioli, E.; Rulli, E.; Rulli, E.; Poli, D.; Legramandi, L. Evaluating the Effects of an Ophthalmic Solution of Coenzyme Q10 and Vitamin E in Open-Angle Glaucoma Patients: A Study Protocol. *Adv. Ther.* **2019**, *36*, 2506–2514. [CrossRef] [PubMed]
99. Verdin, E. NAD^+ in aging, metabolism, and neurodegeneration. *Science* **2015**, *350*, 1208–1213. [CrossRef]
100. Williams, P.A.; Harder, J.M.; Foxworth, N.E.; Cochran, K.E.; Philip, V.M.; Porciatti, V.; Smithies, O.; John, S.W. Vitamin B(3) modulates mitochondrial vulnerability and prevents glaucoma in aged mice. *Science* **2017**, *355*, 756–760. [CrossRef]
101. Kouassi Nzoughet, J.; Chao de la Barca, J.M.; Guehlouz, K.; Leruez, S.; Coulbault, L.; Allouche, S.; Bocca, C.; Muller, J.; Amati-Bonneau, P.; Gohier, P.; et al. Nicotinamide Deficiency in Primary Open-Angle Glaucoma. *Investig. Ophthalmol. Vis. Sci.* **2019**, *60*, 2509–2514. [CrossRef]
102. Hui, F.; Tang, J.; Williams, P.A.; McGuinness, M.B.; Hadoux, X.; Casson, R.J.; Coote, M.; Trounce, I.A.; Martin, K.R.; van Wijngaarden, P.; et al. Improvement in inner retinal function in glaucoma with nicotinamide (vitamin B3) supplementation: A crossover randomized clinical trial. *Clin. Exp. Ophthalmol.* **2020**, *48*, 903–914. [CrossRef] [PubMed]
103. De Moraes, C.G.; John, S.W.M.; Williams, P.A.; Blumberg, D.M.; Cioffi, G.A.; Liebmann, J.M. Nicotinamide and Pyruvate for Neuroenhancement in Open-Angle Glaucoma: A Phase 2 Randomized Clinical Trial. *JAMA Ophthalmol.* **2022**, *140*, 11–18. [CrossRef] [PubMed]
104. Posch-Pertl, L.; Michelitsch, M.; Wagner, G.; Wildner, B.; Silbernagel, G.; Pregartner, G.; Wedrich, A. Cholesterol and glaucoma: A systematic review and meta-analysis. *Acta Ophthalmol.* **2022**, *100*, 148–158. [CrossRef] [PubMed]
105. Kim, M.L.; Sung, K.R.; Kwon, J.; Choi, G.W.; Shin, J.A. Neuroprotective Effect of Statins in a Rat Model of Chronic Ocular Hypertension. *Int. J. Mol. Sci.* **2021**, *22*, 12500. [CrossRef] [PubMed]
106. De Castro, D.K.; Punjabi, O.S.; Bostrom, A.G.; Stamper, R.L.; Lietman, T.M.; Ray, K.; Lin, S.C. Effect of statin drugs and aspirin on progression in open-angle glaucoma suspects using confocal scanning laser ophthalmoscopy. *Clin. Exp. Ophthalmol.* **2007**, *35*, 506–513. [CrossRef]
107. Kang, J.M.; Jammal, A.A.; Medeiros, F.A. Association between statin use and rates of structural and functional loss in glaucoma. *Br. J. Ophthalmol.* **2022**. [CrossRef]
108. Yuan, Y.; Xiong, R.; Wu, Y.; Ha, J.; Wang, W.; Han, X.; He, M. Associations of statin use with the onset and progression of open-angle glaucoma: A systematic review and meta-analysis. *eClinicalMedicine* **2022**, *46*, 101364. [CrossRef]

109. Bibbins-Domingo, K.; Grossman, D.C.; Curry, S.J.; Davidson, K.W.; Epling, J.W., Jr.; García, F.A.R.; Gillman, M.W.; Kemper, A.R.; Krist, A.H.; Kurth, A.E.; et al. Statin Use for the Primary Prevention of Cardiovascular Disease in Adults: US Preventive Services Task Force Recommendation Statement. *JAMA* **2016**, *316*, 1997–2007. [CrossRef]
110. Vergouwen, M.D.; de Haan, R.J.; Vermeulen, M.; Roos, Y.B. Statin treatment and the occurrence of hemorrhagic stroke in patients with a history of cerebrovascular disease. *Stroke* **2008**, *39*, 497–502. [CrossRef]
111. Dahlmann-Noor, A.; Vijay, S.; Jayaram, H.; Limb, A.; Khaw, P.T. Current approaches and future prospects for stem cell rescue and regeneration of the retina and optic nerve. *Can. J. Ophthalmol.* **2010**, *45*, 333–341. [CrossRef]
112. Fu, L.; Kwok, S.S.; Chan, Y.K.; Ming Lai, J.S.; Pan, W.; Nie, L.; Shih, K.C. Therapeutic Strategies for Attenuation of Retinal Ganglion Cell Injury in Optic Neuropathies: Concepts in Translational Research and Therapeutic Implications. *Biomed Res. Int.* **2019**, *2019*, 8397521. [CrossRef] [PubMed]
113. Edo, A.; Sugita, S.; Futatsugi, Y.; Sho, J.; Onishi, A.; Kiuchi, Y.; Takahashi, M. Capacity of Retinal Ganglion Cells Derived from Human Induced Pluripotent Stem Cells to Suppress T-Cells. *Int. J. Mol. Sci.* **2020**, *21*, 7831. [CrossRef]
114. Guo, X.; Zhou, J.; Starr, C.; Mohns, E.J.; Li, Y.; Chen, E.P.; Yoon, Y.; Kellner, C.P.; Tanaka, K.; Wang, H.; et al. Preservation of vision after CaMKII-mediated protection of retinal ganglion cells. *Cell* **2021**, *184*, 4299–4314.e4212. [CrossRef] [PubMed]
115. Harrell, C.R.; Fellabaum, C.; Arsenijevic, A.; Markovic, B.S.; Djonov, V.; Volarevic, V. Therapeutic Potential of Mesenchymal Stem Cells and Their Secretome in the Treatment of Glaucoma. *Stem Cells Int* **2019**, *2019*, 7869130. [CrossRef] [PubMed]
116. Emre, E.; Yüksel, N.; Duruksu, G.; Pirhan, D.; Subaşi, C.; Erman, G.; Karaöz, E. Neuroprotective effects of intravitreally transplanted adipose tissue and bone marrow-derived mesenchymal stem cells in an experimental ocular hypertension model. *Cytotherapy* **2015**, *17*, 543–559. [CrossRef]
117. Harper, M.M.; Grozdanic, S.D.; Blits, B.; Kuehn, M.H.; Zamzow, D.; Buss, J.E.; Kardon, R.H.; Sakaguchi, D.S. Transplantation of BDNF-secreting mesenchymal stem cells provides neuroprotection in chronically hypertensive rat eyes. *Investig. Ophthalmol. Vis. Sci.* **2011**, *52*, 4506–4515. [CrossRef]
118. Johnson, T.V.; Bull, N.D.; Hunt, D.P.; Marina, N.; Tomarev, S.I.; Martin, K.R. Neuroprotective effects of intravitreal mesenchymal stem cell transplantation in experimental glaucoma. *Investig. Ophthalmol. Vis. Sci.* **2010**, *51*, 2051–2059. [CrossRef]
119. Mead, B.; Amaral, J.; Tomarev, S. Mesenchymal Stem Cell-Derived Small Extracellular Vesicles Promote Neuroprotection in Rodent Models of Glaucoma. *Investig. Ophthalmol. Vis. Sci.* **2018**, *59*, 702–714. [CrossRef]
120. Wang, Y.; Lv, J.; Huang, C.; Li, X.; Chen, Y.; Wu, W.; Wu, R. Human Umbilical Cord-Mesenchymal Stem Cells Survive and Migrate within the Vitreous Cavity and Ameliorate Retinal Damage in a Novel Rat Model of Chronic Glaucoma. *Stem Cells Int.* **2021**, *2021*, 8852517. [CrossRef]
121. Roubeix, C.; Godefroy, D.; Mias, C.; Sapienza, A.; Riancho, L.; Degardin, J.; Fradot, V.; Ivkovic, I.; Picaud, S.; Sennlaub, F.; et al. Intraocular pressure reduction and neuroprotection conferred by bone marrow-derived mesenchymal stem cells in an animal model of glaucoma. *Stem Cell Res. Ther.* **2015**, *6*, 177. [CrossRef]
122. Vilela, C.A.P.; Messias, A.; Calado, R.T.; Siqueira, R.C.; Silva, M.J.L.; Covas, D.T.; Paula, J.S. Retinal function after intravitreal injection of autologous bone marrow-derived mesenchymal stromal cells in advanced glaucoma. *Doc. Ophthalmol.* **2021**, *143*, 33–38. [CrossRef] [PubMed]
123. Greco, S.J.; Rameshwar, P. Microenvironmental considerations in the application of human mesenchymal stem cells in regenerative therapies. *Biol. Targets Ther.* **2008**, *2*, 699–705. [CrossRef]
124. Newman, N.J.; Yu-Wai-Man, P.; Carelli, V.; Biousse, V.; Moster, M.L.; Vignal-Clermont, C.; Sergott, R.C.; Klopstock, T.; Sadun, A.A.; Girmens, J.F.; et al. Intravitreal Gene Therapy vs. Natural History in Patients With Leber Hereditary Optic Neuropathy Carrying the m.11778G>A ND4 Mutation: Systematic Review and Indirect Comparison. *Front. Neurol.* **2021**, *12*, 662838. [CrossRef] [PubMed]
125. Wan, X.; Pei, H.; Zhao, M.J.; Yang, S.; Hu, W.K.; He, H.; Ma, S.Q.; Zhang, G.; Dong, X.Y.; Chen, C.; et al. Efficacy and Safety of rAAV2-ND4 Treatment for Leber's Hereditary Optic Neuropathy. *Sci. Rep.* **2016**, *6*, 21587. [CrossRef]
126. Jain, A.; Zode, G.; Kasetti, R.B.; Ran, F.A.; Yan, W.; Sharma, T.P.; Bugge, K.; Searby, C.C.; Fingert, J.H.; Zhang, F.; et al. CRISPR-Cas9-based treatment of myocilin-associated glaucoma. *Proc. Natl. Acad. Sci. USA* **2017**, *114*, 11199–11204. [CrossRef]
127. Osborne, A.; Khatib, T.Z.; Songra, L.; Barber, A.C.; Hall, K.; Kong, G.Y.X.; Widdowson, P.S.; Martin, K.R. Neuroprotection of retinal ganglion cells by a novel gene therapy construct that achieves sustained enhancement of brain-derived neurotrophic factor/tropomyosin-related kinase receptor-B signaling. *Cell Death Dis.* **2018**, *9*, 1007. [CrossRef]
128. Pease, M.E.; Zack, D.J.; Berlinicke, C.; Bloom, K.; Cone, F.; Wang, Y.; Klein, R.L.; Hauswirth, W.W.; Quigley, H.A. Effect of CNTF on retinal ganglion cell survival in experimental glaucoma. *Investig. Ophthalmol. Vis. Sci.* **2009**, *50*, 2194–2200. [CrossRef]
129. Donahue, R.J.; Fehrman, R.L.; Gustafson, J.R.; Nickells, R.W. BCLX(L) gene therapy moderates neuropathology in the DBA/2J mouse model of inherited glaucoma. *Cell Death Dis.* **2021**, *12*, 781. [CrossRef]
130. Fang, F.; Zhuang, P.; Feng, X.; Liu, P.; Liu, D.; Huang, H.; Li, L.; Chen, W.; Liu, L.; Sun, Y.; et al. NMNAT2 is downregulated in glaucomatous RGCs, and RGC-specific gene therapy rescues neurodegeneration and visual function. *Mol. Ther.* **2022**, *30*, 1421–1431. [CrossRef]
131. Lani-Louzada, R.; Marra, C.; Dias, M.S.; de Araújo, V.G.; Abreu, C.A.; Ribas, V.T.; Adesse, D.; Allodi, S.; Chiodo, V.; Hauswirth, W.; et al. Neuroprotective Gene Therapy by Overexpression of the Transcription Factor MAX in Rat Models of Glaucomatous Neurodegeneration. *Investig. Ophthalmol. Vis. Sci.* **2022**, *63*, 5. [CrossRef]
132. Visuvanathan, S.; Baker, A.N.; Lagali, P.S.; Coupland, S.G.; Miller, G.; Hauswirth, W.W.; Tsilfidis, C. XIAP gene therapy effects on retinal ganglion cell structure and function in a mouse model of glaucoma. *Gene Ther.* **2022**, *29*, 147–156. [CrossRef] [PubMed]

Journal of *Personalized Medicine*

Article

Masticatory Muscle Thickness and Activity Correlates to Eyeball Length, Intraocular Pressure, Retinal and Choroidal Thickness in Healthy Women versus Women with Myopia

Grzegorz Zieliński [1,*], Marcin Wójcicki [2], Maria Rapa [3], Anna Matysik-Woźniak [4], Michał Baszczowski [5], Michał Ginszt [6], Monika Litko-Rola [2], Jacek Szkutnik [2], Ingrid Różyło-Kalinowska [7], Robert Rejdak [4] and Piotr Gawda [1]

1 Department of Sports Medicine, Medical University of Lublin, 20-093 Lublin, Poland; piotr.gawda@umlub.pl
2 Independent Unit of Functional Masticatory Disorder, Medical University of Lublin, 20-093 Lublin, Poland; marcin.wojcicki@umlub.pl (M.W.); monikalitkorola@umlub.pl (M.L.-R.); jacek.szkutnik@umlub.pl (J.S.)
3 Students' Scientific Association at the Department and Clinic of General and Pediatric Ophthalmology, Medical University of Lublin, 20-093 Lublin, Poland; maria.rapa00@gmail.com
4 Department of General and Pediatric Ophthalmology, Medical University of Lublin, 20-093 Lublin, Poland; anna.wozniak@umlub.pl (A.M.-W.); robert.rejdak@umlub.pl (R.R.)
5 Interdisciplinary Scientific Group of Sports Medicine, Department of Sports Medicine, Medical University of Lublin, 20-093 Lublin, Poland; m.baszczowski@gmail.com
6 Department of Rehabilitation and Physiotherapy, Medical University of Lublin, 20-093 Lublin, Poland; michal.ginszt@umlub.pl
7 Department of Dental and Maxillofacial Radiodiagnostics with Digital Dentistry Lab, Medical University of Lublin, 20-093 Lublin, Poland; rozylo.kalinowska@umlub.pl
* Correspondence: grzegorz.zielinski@umlub.pl

Citation: Zieliński, G.; Wójcicki, M.; Rapa, M.; Matysik-Woźniak, A.; Baszczowski, M.; Ginszt, M.; Litko-Rola, M.; Szkutnik, J.; Różyło-Kalinowska, I.; Rejdak, R.; et al. Masticatory Muscle Thickness and Activity Correlates to Eyeball Length, Intraocular Pressure, Retinal and Choroidal Thickness in Healthy Women versus Women with Myopia. *J. Pers. Med.* **2022**, *12*, 626. https://doi.org/10.3390/jpm12040626

Academic Editor: Chieh-Chih Tsai

Received: 20 March 2022
Accepted: 11 April 2022
Published: 13 April 2022

Abstract: This study aims to examine the correlations between masticatory and neck muscle thickness and activity versus eyeball length, retinal thickness, choroidal thickness, and intraocular pressure in healthy women versus women with myopia. The study group consisted of 21 women aged 24 years and a control group of 19 women (mean age 23 years). For bioelectrical activity analysis within the temporalis anterior, the superficial part of the masseter muscle, the middle part of the sternocleidomastoid muscle, and the anterior belly of the digastric muscle, an eight-channel BioEMG III electromyograph were used. An M-Turbo ultrasound machine was used to analyze masticatory and neck muscle thickness. The eyeball length was examined by IOL Master 500; choroidal and retinal thickness by Optovue Angiovue; and intraocular pressure by Tono-Pen XL. Refractive errors are related to differences in muscle thickness and electromyographic activity. Bioelectrical activity within the temporalis anterior seems to be associated with ocular length, retinal thickness, and choroidal thickness in women with myopia.

Keywords: eye; optometry; sEMG; eyeball length; retinal; choroidal; intraocular pressure; USG; masticatory muscles; myopia

1. Introduction

The highest rate of increase in myopia is claimed to occur between the ages of 20 and 30. It is estimated that this would affect 72% of individuals and then decrease to 22% in patients over the age of 70. The peak of myopia occurs at the age of 24 [1]. It is predicted that by 2050, there will be 4.8 billion myopic patients [2]. The etiology of myopia is not fully understood and has a multifactorial character. Evidence suggests that not only genetic factors but also environmental factors, such as time spent outdoors, play a major role in the occurrence of myopia [3]. Short-sightedness involving blurring of objects viewed from a distance is related to refractive error, which results from a mismatch between different optical elements of the eye, one of which is the length of the eyeball [4]. The main form of myopia is axial-length myopia, in which the eyeball expands too much. The eyeball length increases during childhood and adolescence, and if this increases the eyeball axial length, it can exceed the focus of the eye leading to myopia [5]. For example, high myopia is defined as a refractive error ≤ -6.00 diopters (D) or an axial length ≥ 26.5 mm [6].

Myopia is known to be associated with several ocular complications such as retinal detachment, glaucoma, cataract, optic nerve disc changes, and maculopathy [7]. These are not the only changes in the structure of the eye associated with myopia. Some studies suggest that intraocular pressure (IOP) is associated with the occurrence of refractive error [8,9]. Based on the Mendel randomization model, it was noted that on average, each 1 mm Hg increase in IOP predicts a decrease in the spherical equivalent of 0.05 to 0.09 Diopters (Dsph) [9]. This suggests that exposure to higher IOP may inadvertently increase the prevalence of myopia [10].

Additionally, there are also changes in the thickness of the retina and choroid in myopia. An overall reduction in retinal thickness has been observed in eyes with myopia compared to eyes without myopia [11–13]. Choroidal thickness in high myopia is inversely correlated with age and myopia refractive error and is an important predictor of visual acuity [14].

An increasing number of studies have recognized changes in the bioelectrical activity of masticatory and cervical spine muscles [15–18]. The connection between the eye and the masticatory muscles may have a neurological, biomechanical, or biochemical basis. Changes in electromyographic activity within masticatory muscles can have a primary or secondary character on the refractive error. However, the presented connection is not yet fully understood [17].

This study aims to examine the correlations between masticatory and neck muscle thickness and activity versus eyeball length, retinal thickness, choroidal thickness, and intraocular pressure in healthy women versus women with myopia. To the best of our knowledge, this is the first study to analyze the relationship between these systems.

2. Materials and Methods

2.1. Study Population

A total of 131 women in the age range of 20–30 years were enrolled in the study. The analysis in this study involved Caucasian women. It was decided to examine women due to the higher prevalence of myopia [19,20], faster progression of myopia in women [21], and higher prevalence of TMD [22,23] compared to men. The following exclusion criteria were used: the presence of TMD based on The Research Diagnostic Criteria for Temporomandibular Disorders (RDC/TMD), open bite, crossbite, Angle class II and III, lack of four spheres of support, oral inflammation, hyperopia, diseases of the optic nerve and ocular structures, neurological disorders of the head and neck, muscular disorders or diseases, neoplastic diseases (regardless of type or location), pain and trauma and previous surgical treatment in the head and neck within the last six months of the study, and pregnancy.

The examination for exclusion criteria was carried out by the authors: a dentist with a specialization in dental prosthetics (author J.S.), a medical doctor specialized in ophthalmology (author A.M.-W.), and a physiotherapist (author G.Z.).

Based on the above criteria, 40 women were included in the study. Refractive error based on ophthalmological examination (author A.M.-W.) was found in 21 subjects. The study group included women with a refractive error from -0.75 to -5.50 Dsph. The ophthalmological examination showed a refractive error in the right eye mean of -3.00 ± 0.50 Dsph and in the left eye mean of -3.00 ± 1.50 Dsph within the study group. The best-corrected visual acuity (BCVA) was tested using a Snellen chart, and it was 1.0 (20/20)). No refractive error was found in the control group (19 women). The uncorrected visual acuity was 1.0 (20/20). Other ophthalmic parameters are listed in Table 1. No other ocular abnormalities were found on a complete ophthalmological examination.

Table 1. Presentation of ophthalmological parameters between the groups.

		Myopia Subject (n = 21)		Subject without Refractive Error (n = 19)				
	Side	**Mean**	**SD**	**Mean**	**SD**	**Test**		p
Refractive Error (Dsph)	R	−3.00	0.50	NA	NA		NA	NA
	L	−3.00	1.50					
Axial Length (mm)	R	24.27	0.69	23.30	0.55	t	4.85	0.00 * ES = 1.56
	L	24.31	0.84	23.27	0.56	t	4.51	0.00 * ES = 1.46
Retinal Thickness (µm)	R	252.71	18.14	244.26	12.16	t	1.71	0.10
	L	245.95	31.10	242.11	11.89	z	1.30	0.19
Choroidal Thickness (µm)	R	349.05	82.86	343.32	82.53	t	0.22	0.83
	L	322.86	83.71	328.89	84.29	t	−0.23	0.82
Intraocular Pressure (mmHg)	R	16.10	4.07	15.11	4.29	t	0.75	0.46
	L	15.48	4.20	14.79	3.85	t	0.54	0.59

NA—not applicable; Dsph—diopters; R—right site; L—left site; mm—millimeters; µm—micrometers; mmHg—millimeter of mercury; t—student t-test; z—Mann–Whitney U test; ES—effect size; *—significant difference.

After the initial measurements and qualification to the study, the study group consisted of 21 women aged 24 years (±3 years) and 19 controls aged 23 years (±2 years). The groups were not statistically significantly different regarding age and number of participants (p = 0.25).

The study was conducted by the recommendations of the 1964 Declaration of Helsinki and its 2013 Seventh Amendment. The study was approved by the local bioethics committee (Bioethics Committee of the Medical University of Lublin, approval no KE-0254/229/2020). All participants were informed of the objectives of the study. They could have withdrawn at any stage of the examination. Each participant provided researchers with their written consent.

2.2. Study Protocol

2.2.1. Assessment of Muscle Thickness

The cross-sectional thickness of the muscles was assessed using an M-Turbo ultrasound machine equipped with a 15–6 MHz linear transducer with scan depth up to 6 cm (SonoSite Inc, Bothell, WA, USA). During the examination, patients were lying down on the dental chair in a horizontal position. The thickness of the masseter muscle was measured with an ultrasound probe positioned perpendicular to the mandibular ramus and in the occlusal plane [24]. The anterior part of the temporalis muscle was examined with a probe positioned in front of the hairline, parallel to and 1 cm above the zygomatic arch, perpendicular to the temporal bone [25]. The anterior belly of the digastric muscle was assessed with a probe positioned in the middle between the hyoid bone and the mental protuberance, perpendicular to the long axis of the muscle and the skin [26]. The thickness of the sternocleidomastoid muscle was examined with a probe positioned perpendicular to the long axis of the muscle and the skin, in the middle between the mastoid process and the clavicular notch of the sternum [27]. The cross-sectional thickness of all the examined muscles on both sides was measured in the relaxed mandible position with slight contact between opposing teeth (Rest) and during maximum voluntary clenching (Clench) (Figure 1). The whole procedure was performed twice for each patient. The thickness was measured in the thickest part of the muscle on the scan, directly on the scanner screen with 0.1 mm accuracy. The mean of the two measurements was included in the analysis. All tests were taken by the same examiner (author M.W.).

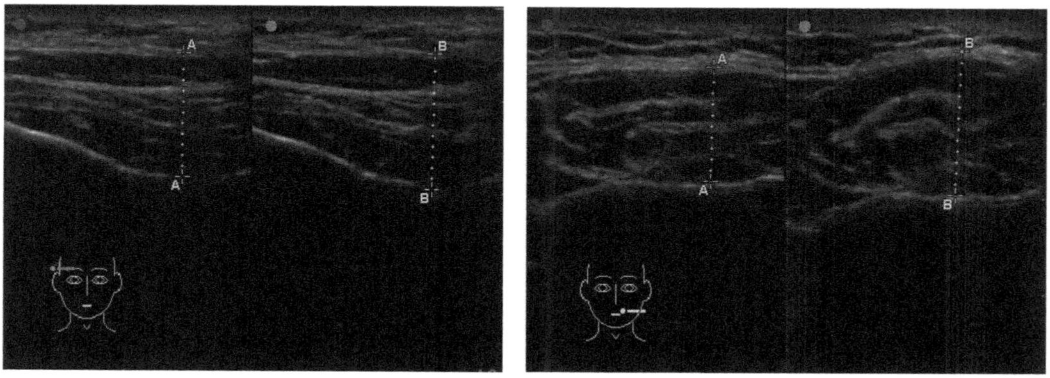

(a) (b) (c) (d)

Figure 1. Example of an ultrasound examination of muscle thickness. (**a**) The temporalis muscle in the relaxed mandible position with slight contact between opposing teeth; (**b**) the temporalis muscle in the maximum voluntary clenching; (**c**) the masseter muscle in the relaxed mandible position with slight contact between opposing teeth; (**d**) the masseter muscle in the maximum voluntary clenching; line A-A—measuring line of the thickness of the muscle in the relaxed mandible position with slight contact between opposing teeth; line B-B—measuring line of the thickness of the maximum voluntary clenching.

2.2.2. Assessment of the Muscle Activity

Masticatory and neck muscle activity was performed using the 8-channel BioEMG III electromyograph compatible with the BioPAK measurement system (BioResearch Associates, Inc., Milwaukee, WI, USA). The subjects took a standardized position in the dental chair [16,28]. The study was conducted between 8 and 12 a.m. The skin of the subjects was cleaned with 90% ethanol, the surface of the electrodes placed on the skin (Ag/AgCl with a diameter of 30 mm and a conductive surface of 16 mm—SORIMEX, Torun, Poland) following the standards of the SENIAM program [29]. Placing electrodes was presented in the earlier study by the authors [16].

Four muscle pairs were analyzed: the anterior part of the temporalis muscle (TA), the superficial part of the masseter muscle (MM), the anterior belly of the digastric muscle (DA), and the middle part of the sternocleidomastoid muscle belly (SCM) [16,28]. Surface electrodes placement and electromyographic (sEMG) examination were taken by the same physiotherapist (author G.Z.).

The electromyography study was performed with the subjects' eyes closed. This was dictated by the exclusion of the possible neurologic component [15–17]. The sEMG activity was recorded during resting mandibular position for 10 s. The sEMG signals obtained during the test were amplified and cleaned according to the previously described methodology [16]. The BioPAK Noise Tests were administered to all participants before and after each measurement. Automatic processing of the sEMG signal based on a root mean square calculation in BioPAK software produced an averaged measure of values, which then were used for muscle-activity analysis (Figure 2).

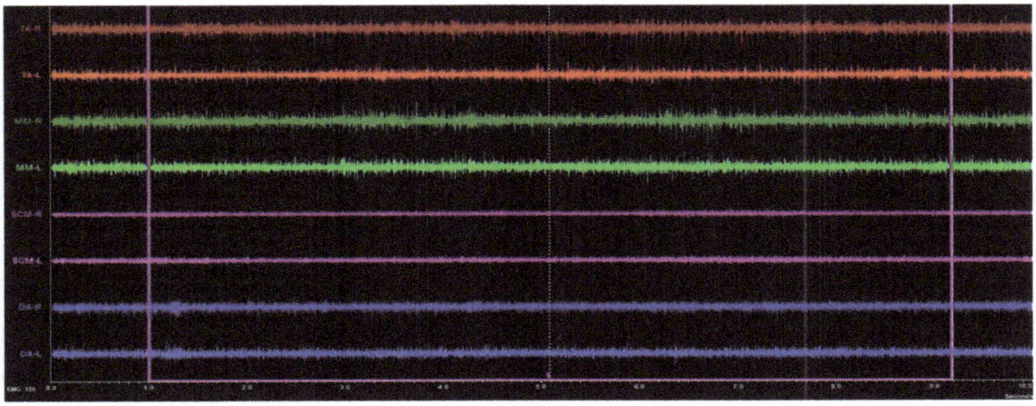

Figure 2. Example of the surface electromyography traces during resting activity.

2.2.3. Ophthalmic Examination

The Axial Length of the Eyeball

An IOL Master 500 (Carl Zeiss Meditec, Jena, Germany) was used to examine eyeball length. It obtains data based on the optical path distance from the anterior surface of the cornea to the retinal pigment epithelium. This device is used in clinical practice to calculate the power of artificial intraocular lenses. The IOL Master 500 is a noninvasive test and uses 780 nm partial coherence interferometry to measure the eye's axial length [30]. All tests were taken by the same examiner (author A.M.-W.). All patients sat in front of the device head, resting their chin and forehead against the tripod. Patients were asked to perform a full blink just before the measurements were taken. This order was given to distribute an optically smooth tear film on the cornea. The eyes were focused when the instrument head was approximately 5.5 cm from the patient. Five separate measurements were taken and averaged for axial length [31].

The Thickness of the Retina and Choroid

Optical coherence tomography (OCT) was used to determine the thickness of the retina and choroid as well as to detect any possible changes in the macula. These measurements are used in a wide range of ophthalmic examinations to determine the thickness, volume, and structure of various retinal layers [32]. Chosen patients sat in front of the device head, resting their chin and forehead against the tripod. The study was performed with Optovue AngioVue (Fremont, CA, USA). To ensure the correctness, the cutoff point for scan quality was 7/10. The retinal and choroidal thickness were measured in the fovea. Area-identification and segmentation methods were automatically selected, and the thickness of the retina was measured in 6 × 6 mm scans. The central macular thickness (CMT) was characterized as the average thickness measured at the point of intersection of the six radial OCT scans. It was automatically measured by the OCT mapping system in the healthy eye. The central foveal thickness (CFT) was defined as the distance between the vitreoretinal interface and the anterior surface of the retinal pigment epithelium (RPE) in the center of fovea; this was measured manually and was also automated using the Optovue AngioVue measurement software. Mean retinal thickness was noted at the central 1 mm. This measurement was given by the automated software (Figure 3A) [33]. The choroidal thickness (CT) was measured manually with the inbuilt caliper in OCT cross-sectional scans. The CT measurement was performed perpendicular to the RPE, going from the posterior RPE edge to the choroid–scleral junction in the center of the fovea (Figure 3B) [33]. The retinal and choroidal thickness measurements were performed at the same time of day between 1 and 3 p.m. to eliminate the changes caused by the time of day. Evaluation of

OCT scans and measurement of retinal and choroidal thickness were performed by two independent readers (A.M.-W. and M.R.) [32,34] (Figure 3).

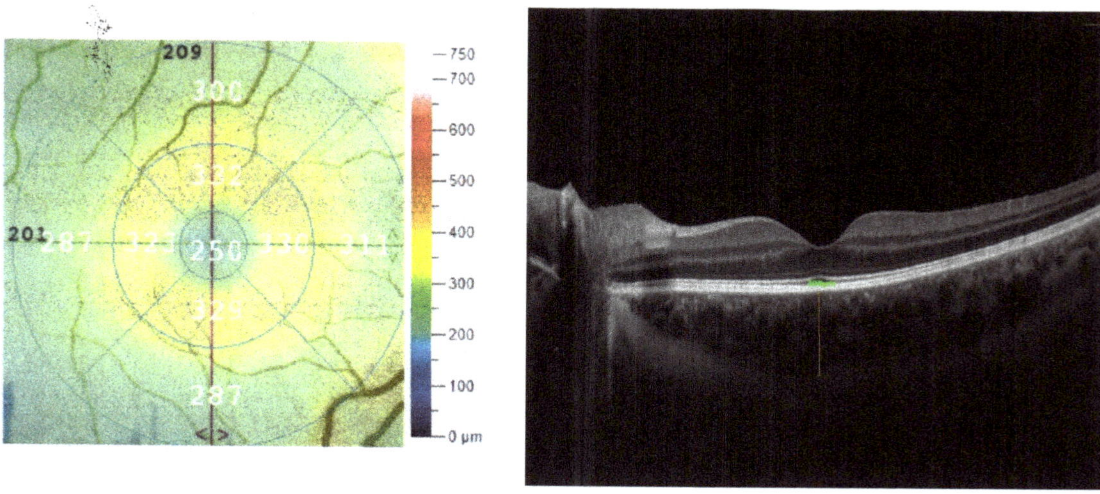

(**A**) (**B**)

Figure 3. Example of a study of retinal thickness (**A**) and choroidal thickness (**B**).

The Intraocular Pressure

The final test was taken to determine the intraocular pressure. It was performed at the end of the whole examination so that the administered anesthesia and potential epithelial defects would not affect other measurements. The intraocular pressure was measured with a Tono-Pen XL (Medtronic Solan, FL, USA) in both eyes after the application of ALCAINE 0.5% (Alcon Laboratories Inc., Fort Worth, TX, USA). The apparatus was positioned vertically to the anesthetized cornea. The device uses voltage micrometer technology and a 1.0 mm transducer tip. The tip of the tonometer compresses the cornea, and its resistance depends on the pressure in the eyeball. The Tono-Pen XL gently shrinks the cornea and displays an average of four independent readings and a statistical factor [35,36].

2.2.4. Statistical Analysis

For statistical analysis, IBM SPSS Statistics 13.3 program was used. First, the Shapiro–Wilk test and the Kolmogorov–Smirnov test (with Lillierfors correction) were applied to check the distribution. The Student's *t*-test (t) or Mann–Whitney U test (z) was used to compare the differences between ophthalmological parameters, bioelectrical tension, and muscle thicknesses (TA, MM, SCM, and DA) in the study groups, depending on the distribution. The Spearman rho test (s) and r-Pearson test (p) (depending on the distribution) were used to analyze the correlation between the length of the eyeball, reticular thickness, choroidal thickness, intraocular pressure, and the thickness and bioelectrical tension of selected muscles of the masticatory system and the cervical segment. The Spearman rho test (s) and r-Pearson test varied between −1 (perfect negative monotonic association) and +1 (perfect positive monotonic association). A correlation was considered large for values greater than 0.5 and moderate for values between 0.3 and 0.5 [37]. Effect sizes were determined for the *t*-test using the Cohen d method and interpreted as small (0.2), medium (0.5), and large (0.8) effect sizes [38,39]. Statistical significance was set at $p \leq 0.05$.

3. Results

The groups differed significantly in the axial length of the eyeball. The other parameters did not differ statistically (Table 1).

The results of the between-group analysis showed significant differences between TA Rest $_R$ muscle thickness in clenching. A larger cross-sectional TA $_R$ was observed in subjects with myopia. Moreover, a smaller SCM Rest $_R$ and SCM Clench $_R$ thickness was also observed in subjects with myopia in comparison to women without the refractive error (Table 2).

Table 2. Comparison of masticatory and cervical muscle thickness and activity between groups.

	Myopia Subject (n = 21)		Subject without Refractive Error (n = 19)		Test		p
	Mean	SD	Mean	SD			
TA Rest$_{R\ (mm)}$	13.18	1.78	12.20	1.48	t	1.90	0.06
TA Clench$_{R\ (mm)}$	13.86	1.82	12.77	1.49	t	2.06	0.04 * ES = 0.60
MM Rest$_{R\ (mm)}$	11.94	2.18	12.42	1.51	t	−0.81	0.42
MM Clench$_{R\ (mm)}$	13.48	2.39	13.52	1.58	t	−0.05	0.96
DA Rest$_{R\ (mm)}$	6.55	0.63	6.67	0.81	z	−0.33	0.75
DA Clench$_{R\ (mm)}$	6.47	0.55	6.67	0.83	z	−0.64	0.52
SCM Rest$_{R\ (mm)}$	9.36	1.37	10.25	1.37	t	−2.05	0.04 * ES = 0.65
SCM Clench$_{R\ (mm)}$	9.37	1.31	10.26	1.42	t	−2.06	0.04 * ES = 0.69
TA$_{R\ (\mu V)}$	1.80	0.84	2.05	1.99	z	0.66	0.51
MM$_{R\ (\mu V)}$	2.53	1.67	2.18	1.31	z	0.58	0.56
DA$_{R\ (\mu V)}$	1.30	0.37	1.16	0.39	z	−0.96	0.34
SCM$_{R\ (\mu V)}$	1.70	0.75	1.93	0.76	z	1.57	0.12
TA Rest$_{L\ (mm)}$	12.25	1.78	12.20	1.90	z	−0.03	0.98
TA Clench$_{L\ (mm)}$	12.88	1.87	12.80	2.00	z	0.22	0.83
MM Rest$_{L\ (mm)}$	12.43	2.30	12.69	1.55	z	−1.02	0.31
MM Clench$_{L\ (mm)}$	13.78	2.40	13.75	1.53	z	−0.03	0.98
DA Rest$_{L\ (mm)}$	6.70	0.97	6.56	0.75	t	0.37	0.71
DA Clench$_{L\ (mm)}$	6.61	0.98	6.58	0.74	t	0.09	0.93
SCM Rest$_{L\ (mm)}$	9.51	1.55	10.00	1.14	t	−1.12	0.27
SCM Clench$_{L\ (mm)}$	9.56	1.55	9.99	1.15	t	−1.00	0.33
TA$_{L\ (\mu V)}$	1.90	0.93	2.30	1.80	z	−0.35	0.72
MM$_{L\ (\mu V)}$	2.01	1.09	2.40	1.65	z	−1.14	0.26
DA$_{L\ (\mu V)}$	1.40	0.49	1.33	0.45	z	−0.22	0.83
SCM$_{L\ (\mu V)}$	1.71	0.75	1.63	0.54	z	0.50	0.62

TA—the temporalis anterior; MM—the superficial part of the masseter muscle; SCM—the middle part of the sternocleidomastoid muscle; DA—the anterior belly of the digastric muscle; $_R$—right site; $_L$—left site; $_{\mu V}$—microvolt; $_{mm}$—millimeters; Rest—ultrasound muscles examination in the relaxed mandible position with slight contact between opposing teeth; Clench—ultrasound muscles examination in the maximum voluntary clenching; t—student *t*-test; z—Mann–Whitney U test; ES—effect size; *—significant difference.

Correlation (right-hand) showed a negative correspondence between TA $_R$ bioelectrical voltage and retinal thickness in the myopia group. In the group without refractive error, co-correlations were observed between ocular length and MM Rest $_R$ and MM Clench $_R$, retinal thickness and SCM Rest $_R$ and SCM Clench, along with intraocular pressure and SCM $_R$ bioelectrical activity (Table 3).

Table 3. Presentation of the resulting right-hand correlations between groups.

		Myopia Subject		Subject without Refractive Error	
		r	p	r	p
Axial Length$_R$ (mm)	TA Rest$_{R\,(mm)}$	0.18 P	0.43	0.33 P	0.17
	TA Clench$_{R\,(mm)}$	0.24 P	0.30	0.32 P	0.19
	MM Rest$_{R\,(mm)}$	0.08 P	0.72	0.52 P	0.02 *
	MM Clench$_{R\,(mm)}$	0.06 P	0.80	0.62 P	0.00 *
	DA Rest$_{R\,(mm)}$	0.28 s	0.21	−0.05 P	0.85
	DA Clench$_{R\,(mm)}$	0.18 s	0.43	−0.03 P	0.92
	SCM Rest$_{R\,(mm)}$	−0.23 P	0.31	−0.28 P	0.24
	SCM Clench$_{R\,(mm)}$	−0.23 P	0.32	−0.30 P	0.21
	TA$_{R\,(\mu V)}$	−0.24 s	0.29	0.14 s	0.57
	MM$_{R\,(\mu V)}$	0.34 s	0.13	−0.03 s	0.90
	DA$_{R\,(\mu V)}$	−0.23 P	0.32	0.20 s	0.41
	SCM$_{R\,(\mu V)}$	0.29 P	0.20	0.23 s	0.35
Retinal Thickness R (µm)	TA Rest$_{R\,(mm)}$	0.40 P	0.07	−0.15 P	0.53
	TA Clench$_{R\,(mm)}$	0.38 P	0.09	−0.24 P	0.32
	MM Rest$_{R\,(mm)}$	−0.28 P	0.21	−0.17 P	0.49
	MM Clench$_{R\,(mm)}$	−0.09 P	0.69	−0.06 P	0.82
	DA Rest$_{R\,(mm)}$	−0.09 s	0.71	−0.03 P	0.89
	DA Clench$_{R\,(mm)}$	−0.13 s	0.59	−0.07 P	0.76
	SCM Rest$_{R\,(mm)}$	−0.26 P	0.25	−0.47 P	0.04 *
	SCM Clench$_{R\,(mm)}$	−0.25 P	0.27	−0.48 P	0.04 *
	TA$_{R\,(\mu V)}$	−0.52 s	0.02 *	−0.21 s	0.39
	MM$_{R\,(\mu V)}$	−0.23 s	0.31	0.12 s	0.62
	DA$_{R\,(\mu V)}$	−0.28 P	0.22	0.36 s	0.13
	SCM$_{R\,(\mu V)}$	−0.03 P	0.91	0.07 s	0.77
Choroidal Thickness R (µm)	TA Rest$_{R\,(mm)}$	0.38 P	0.09	−0.17 P	0.48
	TA Clench$_{R\,(mm)}$	0.37 P	0.10	−0.16 P	0.51
	MM Rest$_{R\,(mm)}$	−0.08 P	0.72	−0.33 P	0.17
	MM Clench$_{R\,(mm)}$	0.02 P	0.92	−0.22 P	0.36
	DA Rest$_{R\,(mm)}$	−0.19 s	0.41	−0.13 P	0.59
	DA Clench$_{R\,(mm)}$	−0.28 s	0.23	−0.14 P	0.56
	SCM Rest$_{R\,(mm)}$	−0.09 P	0.69	0.18 P	0.46
	SCM Clench$_{R\,(mm)}$	−0.13 P	0.57	0.19 P	0.44
	TA$_{R\,(\mu V)}$	0.29 s	0.21	−0.28 s	0.24
	MM$_{R\,(\mu V)}$	0.19 s	0.41	−0.10 s	0.69
	DA$_{R\,(\mu V)}$	0.10 P	0.66	0.08 s	0.74
	SCM$_{R\,(\mu V)}$	0.37 P	0.09	−0.21 s	0.39
Intraocular Pressure$_R$ (mmHg)	TA Rest$_{R\,(mm)}$	0.16 P	0.49	0.37 P	0.11
	TA Clench$_{R\,(mm)}$	0.18 P	0.43	0.34 P	0.15
	MM Rest$_{R\,(mm)}$	−0.04 P	0.88	−0.07 P	0.78
	MM Clench$_{R\,(mm)}$	0.09 P	0.70	−0.08 P	0.74
	DA Rest$_{R\,(mm)}$	0.36 s	0.11	−0.14 P	0.58
	DA Clench$_{R\,(mm)}$	0.40 s	0.08	−0.10 P	0.69
	SCM Rest$_{R\,(mm)}$	−0.23 P	0.31	−0.14 P	0.57
	SCM Clench$_{R\,(mm)}$	−0.23 P	0.31	−0.15 P	0.55
	TA$_{R\,(\mu V)}$	0.33 s	0.14	0.10 s	0.68
	MM$_{R\,(\mu V)}$	0.08 s	0.74	−0.31 s	0.19
	DA$_{R\,(\mu V)}$	0.17 P	0.46	0.07 s	0.78
	SCM$_{R\,(\mu V)}$	−0.20 P	0.38	−0.49 s	0.03 *

TA—the temporalis anterior; MM—the superficial part of the masseter muscle; SCM—the middle part of the sternocleidomastoid muscle; DA—the anterior belly of the digastric muscle; $_R$—right site; $_{\mu V}$—microvolt; $_{mm}$—millimeters; $_{\mu m}$—micrometers; $_{mmHg}$—millimeter of mercury; Rest—ultrasound muscles examination in the relaxed mandible position with slight contact between opposing teeth; Clench—ultrasound muscles examination in the maximum voluntary clenching; s—the Spearman rho test; P—the Pearson test; ES—effect size; *—significant difference.

Correlation (left-hand) showed a correspondence between eyeball length and TA L bioelectric active, between choroidal thickness and TA L and SCM L bioelectric active, between intraocular pressure and DA Clench L in the myopia group. No left-hand correlations were observed in the no-refractive-error group (Table 4).

Table 4. Presentation of the resulting left-hand correlations between groups.

		Myopia Subject		**Subject without Refractive Error**	
		r	*p*	r	*p*
Axial Length$_L$ (mm)	TA Rest$_{L\ (mm)}$	−0.08 ^{P}	0.72	0.28 ^{S}	0.25
	TA Clench$_{L\ (mm)}$	0.18 ^{P}	0.43	0.32 ^{S}	0.19
	MM Rest$_{L\ (mm)}$	0.27 ^{S}	0.23	0.21 ^{P}	0.39
	MM Clench$_{L\ (mm)}$	0.19 ^{P}	0.41	0.38 ^{P}	0.10
	DA Rest$_{L\ (mm)}$	0.12 ^{P}	0.60	−0.22 ^{S}	0.37
	DA Clench$_{L\ (mm)}$	0.06 ^{P}	0.79	−0.30 ^{P}	0.21
	SCM Rest$_{L\ (mm)}$	−0.23 ^{P}	0.33	0.03 ^{P}	0.91
	SCM Clench$_{L\ (mm)}$	−0.13 ^{S}	0.58	0.02 ^{P}	0.94
	TA$_{L\ (\mu V)}$	−0.76 ^{S}	0.00 *	0.00 ^{S}	1.00
	MM$_{L\ (\mu V)}$	0.03 ^{S}	0.91	0.21 ^{S}	0.38
	DA$_{L\ (\mu V)}$	−0.12 ^{S}	0.61	0.43 ^{P}	0.07
	SCM$_{L\ (\mu V)}$	−0.37 ^{S}	0.10	0.36 ^{S}	0.13
Retinal Thickness$_L$ (µm)	TA Rest$_{L\ (mm)}$	−0.16 ^{S}	0.48	−0.06 ^{S}	0.80
	TA Clench$_{L\ (mm)}$	−0.09 ^{S}	0.68	−0.03 ^{S}	0.92
	MM Rest$_{L\ (mm)}$	0.21 ^{S}	0.37	0.00 ^{P}	1.00
	MM Clench$_{L\ (mm)}$	0.10 ^{S}	0.68	0.03 ^{P}	0.89
	DA Rest$_{L\ (mm)}$	0.07 ^{S}	0.76	0.04 ^{S}	0.86
	DA Clench$_{L\ (mm)}$	−0.05 ^{S}	0.83	0.01 ^{P}	0.98
	SCM Rest$_{L\ (mm)}$	−0.15 ^{S}	0.53	−0.10 ^{P}	0.68
	SCM Clench$_{L\ (mm)}$	−0.12 ^{S}	0.59	−0.11 ^{P}	0.65
	TA$_{L\ (\mu V)}$	−0.26 ^{S}	0.25	−0.11 ^{S}	0.66
	MM$_{L\ (\mu V)}$	−0.17 ^{S}	0.45	−0.05 ^{S}	0.85
	DA$_{L\ (\mu V)}$	0.07 ^{S}	0.77	0.39 ^{P}	0.09
	SCM$_{L\ (\mu V)}$	−0.12 ^{S}	0.60	0.00 ^{S}	0.99
Choroidal thickness$_L$ (µm)	TA Rest$_{L\ (mm)}$	0.06 ^{P}	0.80	0.17 ^{S}	0.48
	TA Clench$_{L\ (mm)}$	−0.15 ^{P}	0.52	0.09 ^{S}	0.73
	MM Rest$_{L\ (mm)}$	−0.19 ^{S}	0.40	−0.31 ^{P}	0.20
	MM Clench$_{L\ (mm)}$	−0.21 ^{P}	0.36	−0.20 ^{P}	0.41
	DA Rest$_{L\ (mm)}$	−0.18 ^{P}	0.42	−0.16 ^{S}	0.50
	DA Clench$_{L\ (mm)}$	0.04 ^{P}	0.87	−0.22 ^{P}	0.36
	SCM Rest$_{L\ (mm)}$	−0.01 ^{P}	0.97	0.28 ^{P}	0.24
	SCM Clench$_{L\ (mm)}$	−0.04 ^{S}	0.88	0.29 ^{P}	0.23
	TA$_{L\ (\mu V)}$	0.53 ^{S}	0.01 *	−0.27 ^{S}	0.26
	MM$_{L\ (\mu V)}$	0.09 ^{S}	0.71	0.08 ^{S}	0.75
	DA$_{L\ (\mu V)}$	0.06 ^{S}	0.78	0.12	0.63
	SCM$_{L\ (\mu V)}$	0.46 ^{S}	0.04 *	0.20 ^{S}	0.42

Table 4. *Cont.*

		Myopia Subject		Subject without Refractive Error	
		r	*p*	r	*p*
Intraocular Pressure L (mmHg)	TA Rest$_{L\ (mm)}$	0.35 P	0.12	0.09 s	0.71
	TA Clench$_{L\ (mm)}$	0.29 P	0.20	0.09 s	0.71
	MM Rest$_{L\ (mm)}$	−0.10 s	0.66	0.11 P	0.65
	MM Clench$_{L\ (mm)}$	0.08 P	0.73	0.13 P	0.58
	DA Rest$_{L\ (mm)}$	0.43 P	0.05	0.39 s	0.10
	DA Clench$_{L\ (mm)}$	0.48 P	0.03 *	0.36 P	0.14
	SCM Rest$_{L\ (mm)}$	−0.13 P	0.58	0.08 P	0.75
	SCM Clench$_{L\ (mm)}$	−0.24 s	0.30	0.07 P	0.77
	TA$_{L\ (\mu V)}$	0.22 s	0.33	−0.11 s	0.65
	MM$_{L\ (\mu V)}$	−0.04 s	0.88	−0.08 s	0.73
	DA$_{L\ (\mu V)}$	−0.31 s	0.17	0.06 P	0.79
	SCM$_{L\ (\mu V)}$	0.16 s	0.50	−0.17 s	0.48

TA—the temporalis anterior; MM—the superficial part of the masseter muscle; SCM—the middle part of the sternocleidomastoid muscle; DA—the anterior belly of the digastric muscle; L—left site; μV—microvolt; mm—millimeters; μm—micrometers, mmHg—millimeter of mercury; Rest—ultrasound muscles examination in the relaxed mandible position with slight contact between opposing teeth; Clench—ultrasound muscles examination in the maximum voluntary clenching; s—the Spearman rho test; P—the Pearson test; ES—effect size; *—significant difference.

4. Discussion

This study is aimed to examine the correlations between masticatory and neck muscle thickness and activity versus eyeball length, retinal thickness, choroidal thickness, and intraocular pressure in healthy women versus women with myopia.

Subjects with myopia had lower SCM thickness compared to subjects without refractive error. In myopic subjects, single correlations were seen between eyeball length, retinal thickness, choroidal thickness, intraocular pressure, and masticatory muscle bioelectrical activity. There was a single correlation between muscle thickness during contraction and intraocular pressure, and it was on the border of a medium-impact effect. In the group without refractive error, correlations were observed with MM thickness versus ocular depth length and SCM thickness, retinal thickness, and intraocular pressure versus SCM bioelectrical activity.

The results obtained during the procedure confirm the interdependence between the stomatognathic system and the organ of vision. The analysis was performed with closed eyes in the test and control groups. This allowed the exclusion of the neurological component [15–17], and therefore, the results should be considered as functional/biomechanical or biochemical changes.

The standard for the axial length of the eyeball in healthy subjects is assumed to be about 23.75 mm [40,41]. The axial length of the eyeball is 24 mm for low myopia (−6 D < refractive error < 0 D), whereas the eyeball length for high myopia is approximately 30 mm (refractive error < −6 D) [42]. The study group included subjects with axial myopia, and their eyeball length was above 24 mm (Table 1). The retinal thickness of both eyes in the study group was greater than in the control group. According to a study by Zereid and Osuagwu, the retinal thickness measured in the fovea in healthy subjects is 238.5 ± 8.4, and in subjects with low myopia is 253.4 ± 8.4 at a mean age of 27 years [43]. Greater retinal thickness in subjects with myopia compared to those without refractive error was also recognized in the author's study but did not reach the assumed level of significance. (Table 1). The observed thicknesses were also similar. In healthy subjects, the normal choroidal thickness is about 250 to 350 μm [44]. According to Hoseini-Yazdi's studies, the thickness of the choroid changes with the refractive error; myopia subjects had a thinner choroid than people without refractive error [45]. This was seen unilaterally in the author's study (the left choroid was thinner) compared to the treatment group. According to the research of Gupta et al., mean wake-up pressure was defined as 16.12 ± 2.94 for the group with myopia and 15.26 ± 2.74 for the group without refractive error [46]. Similar values were achieved in the author's study of higher intraocular pressure in patients with myopia (Table 1). The above differences observed in the author's study and the studies of other authors indicate significant changes in the subjects whose eyes did not have refractive error compared to the one with myopia. The main consideration should be whether the changes in the visual system are primary or secondary compared to the changes in the myofascial system. Changes in the axial length of the eyeball, retinal thickness, choroidal thickness, and intraocular pressure may indicate a mechanical component [17].

The connection between the visual organ and the muscles of the chewing organ can take place via a fascia network [17]. The Tenon's fascia begins at the optic nerve sheath deep inside the orbit. It divides in this area into two parts: the first one extends anteriorly surrounding the eyeball up to the oculomotor attachment and further connects with the conjunctiva [47]. The second part forms on the meatus of the oculomotor muscles. It further connects to the orbital periosteum and the fascia of the eyelid elevator. This system of connections lets the eyeball see the appropriate shape [48]. The Tenon's fascia surrounds six oculomotor muscles and connects with the fascia of the eyelid lever muscle [47,49]. It is not possible to separate the function of the eye muscle from the action of the Tenon fascia [47]. For three-dimensional vision to occur, the eye muscles must constantly be active and perform movements to adjust the visual organ to the objects to be observed [48]. The correct function of the optic organ and the axial length of the eyeball, the thickness of the retina, the structure of the choroid, and the intraocular pressure may depend strongly on the external and intraocular musculature. Muscle tension and its elasticity are related to fascia [17,48]. Tenon's fascia surrounds six oculomotor muscles and connects with the fascia of the eyelid lever muscle [47,49]. The oculomotor belongs to the superficial musculoaponeurotic system (SMAS), including the facial muscle [50]. SMAS is combined with epicranial aponeurosis and successively with the cervical ligament. At the front, the SMAS connects with the superficial fascia surrounding the platysma muscle [47,50]. The temporal parietal fascia connects with the SMAS [51]. These connections can explain changes in the TA, MM, SCM, and DA muscles. (Figure 4).

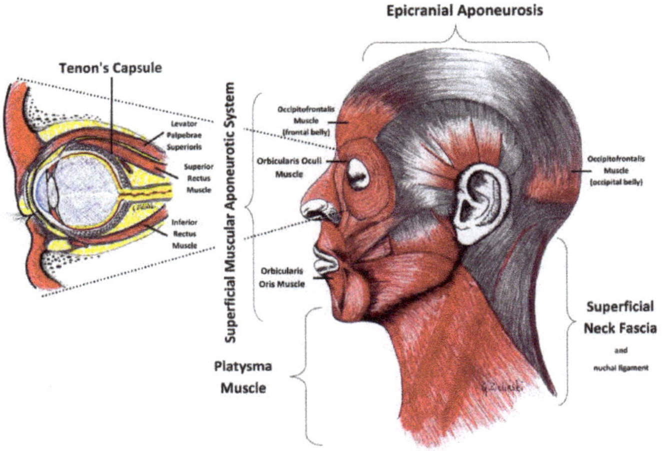

Figure 4. Scheme of anatomical compounds.

The difference in TA thickness between the two groups may be associated with greater activation of these muscles in myopic subjects. What is also worth stressing is that uncorrected refractive errors very frequently coexist with headaches [52]. According to Marasini et al., refractive and binocular vision abnormalities should be investigated in detail in all patients with headaches [52]. One of the causes of headaches may be the activity of TA and increased tension in the epicranial aponeurosis [53,54]. The statistically significant differences between groups in SCM thickness may be related to greater SCM muscle activity in individuals with myopia. Patients often compensate for visual problems by leaning towards the object or turning their heads from side to side. Moreover, individuals with myopia often protract the head and cervical spine, leading to increased tension in the thoracic muscles, descending fibers of the quadriceps, scapular elevator, and sternocleidomastoid muscles [55]. The head position and mandibular movements (lateral flexion and tilt) have been perceived as the ones that affect SCM thickness and activity, especially in the clenched state [56]. Headaches associated with increased TA activity, epicranial aponeurosis, and a change in protracted alignment may be related to changes in proprioception. The cervical proprioceptive system (CPS) consists of mechanoreceptors located in the deep cervical fascia, ligaments, short cervical spine muscles, and intervertebral joints [57]. Information from the cervical region is combined with information received from the vestibular bite [58,59]. The activity of the vestibular bite in people with nearsightedness is specifically linked to the vestibulo-ocular reflex (VOR) [15,17]. People with myopia may experience changes in their perception of visual data. These inputs play an important role in the maintenance and modifications of muscle basal tone [60,61]. Afferent impulses from proprioceptors cooperate with labyrinthine impulses to support oculomotor muscle activity through the VOR [15]. The connection between the cervical region and changes in masticatory muscle activity has been confirmed [62]. It is worth considering whether the changes in the cervical area and tension of facial and cranial muscles are primary or secondary to the changes in the visual system. The influence of myopia on TA, MM, SCM, and DA activity is still not fully understood. An anatomical connection is undoubtedly discernible.

In subjects without refractive error, only unilateral correlations related to MM thickness compared to eyeball length and SCM versus retinal thickness were noted. The unilateral correlations may be related to a preference for the masticatory side [63,64]. Unilateral stimulation could also transmit tensions via the above-mentioned fascial pathway. The phenomenon needs further study.

The presented study has several limitations. Firstly, the group is homogeneous and should be expanded to include the male population. Secondly, the size of the examined group was small. Therefore, future studies should be prospectively examined on a larger age-diverse population. Moreover, refractive errors are also linked to race [65,66]. The analysis in this study involved Caucasian women. Prospective studies should be conducted on different races to compare their link. Thirdly, in the presented study, we used RDC/TMD criteria. Diagnostic criteria for TMDs were changed to The Diagnostic Criteria for Temporomandibular Disorders (DC/TMDs) in 2014. So far, DC/TMDs have not been translated into Polish.

5. Conclusions

Refractive errors are related to differences in masticatory and neck muscle thickness and activity. Bioelectrical activity within the temporalis anterior seems to be associated with ocular length, retinal thickness, and choroidal thickness in women with myopia.

Author Contributions: Conceptualization, G.Z., A.M.-W. and P.G.; methodology, G.Z., M.W., A.M.-W., M.G. and P.G.; formal analysis, G.Z., M.W., A.M.-W., M.G., J.S. and P.G.; investigation, G.Z., M.W., M.R., A.M.-W., M.B., M.G., M.L.-R. and J.S.; resources, G.Z., M.R., A.M.-W., M.B., M.G., M.L.-R. and J.S.; data curation, G.Z., M.W. and M.G.; writing—original draft preparation, G.Z., M.G. and M.W.; writing—review and editing, G.Z., M.R., A.M.-W., M.B. and M.G.; visualization, G.Z. and M.W.; supervision, A.M.-W., I.R.-K., R.R. and P.G.; project administration, G.Z.; funding acquisition, G.Z. All authors have read and agreed to the published version of the manuscript.

Funding: This research received no external funding.

Institutional Review Board Statement: The study was conducted according to the guidelines of the Declaration of Helsinki and approved by the Bioethics Committee of the Medical University of Lublin (approval number KE-0254/229/2020).

Informed Consent Statement: Informed consent was obtained from all subjects involved in the study. Written informed consent has been obtained from the patients to publish this paper.

Data Availability Statement: The datasets generated during and/or analyzed during the current study are available from the corresponding author on reasonable request.

Acknowledgments: We would like to thank all the participants. We acknowledge Julia Różycka for the help with preparing the final version of the manuscript.

Conflicts of Interest: The authors declare no conflict of interest.

Abbreviations

μm	micrometers
μV	microvolt
CFT	central foveal thickness
Clench	ultrasound muscles examination in the maximum voluntary clenching
CPS	cervical proprioceptive system
DA	anterior belly of the digastric muscle
Dsph	Diopters
IOP	intraocular pressure
L	Left site
mm	millimeters
MM	the superficial part of the masseter muscle
mmHg	millimeter of mercury
P	Pearson test
R	Right site
RDC/TMD	Research Diagnostic Criteria for Temporomandibular Disorders
Rest	ultrasound muscles examination in the relaxed mandible position with slight contact between opposing teeth
s	Spearman rho test
SCM	middle part of the sternocleidomastoid muscle belly
sEMG	surface electromyography
SMAS	superficial musculoaponeurotic system
TA	part of the temporalis muscle
TMD	temporomandibular disorders
VOR	vestibulo-ocular reflex

References

1. Hrynchak, P.K.; Mittelstaedt, A.; Machan, C.M.; Bunn, C.; Irving, E.L. Increase in Myopia Prevalence in Clinic-Based Populations across a Century. *Optom. Vis. Sci.* **2013**, *90*, 1331–1341. [CrossRef]
2. Holden, B.A.; Fricke, T.R.; Wilson, D.A.; Jong, M.; Naidoo, K.S.; Sankaridurg, P.; Wong, T.Y.; Naduvilath, T.J.; Resnikoff, S. Global Prevalence of Myopia and High Myopia and Temporal Trends from 2000 through 2050. *Ophthalmology* **2016**, *123*, 1036–1042. [CrossRef]
3. Grzybowski, A.; Kanclerz, P.; Tsubota, K.; Lanca, C.; Saw, S.-M. A Review on the Epidemiology of Myopia in School Children Worldwide. *BMC Ophthalmol.* **2020**, *20*, 27. [CrossRef] [PubMed]
4. Carr, B.J.; Stell, W.K. The Science Behind Myopia. In *Webvision: The Organization of the Retina and Visual System*; Kolb, H., Fernandez, E., Nelson, R., Eds.; University of Utah Health Sciences Center: Salt Lake City, UT, USA, 1995.
5. Tideman, J.W.L.; Polling, J.R.; Vingerling, J.R.; Jaddoe, V.W.V.; Williams, C.; Guggenheim, J.A.; Klaver, C.C.W. Axial Length Growth and the Risk of Developing Myopia in European Children. *Acta Ophthalmol.* **2018**, *96*, 301–309. [CrossRef]
6. Flitcroft, D.I.; He, M.; Jonas, J.B.; Jong, M.; Naidoo, K.; Ohno-Matsui, K.; Rahi, J.; Resnikoff, S.; Vitale, S.; Yannuzzi, L. IMI—Defining and Classifying Myopia: A Proposed Set of Standards for Clinical and Epidemiologic Studies. *Investig. Ophthalmol. Vis. Sci.* **2019**, *60*, M20–M30. [CrossRef] [PubMed]
7. Ikuno, Y. Overview of the complications of high myopia. *Retina* **2017**, *37*, 2347–2351. [CrossRef]
8. Manny, R.E.; Mitchell, G.L.; Cotter, S.A.; Jones-Jordan, L.A.; Kleinstein, R.N.; Mutti, D.O.; Twelker, J.D.; Zadnik, K. CLEERE Study Group Intraocular Pressure, Ethnicity, and Refractive Error. *Optom. Vis. Sci.* **2011**, *88*, 1445–1453. [CrossRef]
9. Hysi, P.G.; Choquet, H.; Khawaja, A.P.; Wojciechowski, R.; Tedja, M.S.; Yin, J.; Simcoe, M.J.; Patasova, K.; Mahroo, O.A.; Thai, K.K.; et al. Meta-Analysis of 542,934 Subjects of European Ancestry Identifies New Genes and Mechanisms Predisposing to Refractive Error and Myopia. *Nat. Genet.* **2020**, *52*, 401–407. [CrossRef] [PubMed]

10. Wang, P.; Chen, S.; Liu, Y.; Lin, F.; Song, Y.; Li, T.; Aung, T.; Zhang, X. Lowering Intraocular Pressure: A Potential Approach for Controlling High Myopia Progression. *Investig. Ophthalmol. Vis. Sci.* **2021**, *62*, 17. [CrossRef] [PubMed]
11. Cheng, S.C.K.; Lam, C.S.Y.; Yap, M.K.H. Retinal Thickness in Myopic and Non-Myopic Eyes. *Ophthalmic Physiol. Opt.* **2010**, *30*, 776–784. [CrossRef]
12. Myers, C.E.; Klein, B.E.K.; Meuer, S.M.; Swift, M.K.; Chandler, C.S.; Huang, Y.; Gangaputra, S.; Pak, J.W.; Danis, R.P.; Klein, R. Retinal Thickness Measured by Spectral Domain Optical Coherence Tomography in Eyes without Retinal Abnormalities: The Beaver Dam Eye Study. *Am. J. Ophthalmol.* **2015**, *159*, 445–456.e1. [CrossRef] [PubMed]
13. Liu, C.; Wei, P.; Li, J. The Thickness Changes of Retina in High Myopia Patients during the Third Trimester of Pregnancy: A Pilot Study. *BMC Ophthalmol.* **2021**, *21*, 382. [CrossRef] [PubMed]
14. Nishida, Y.; Fujiwara, T.; Imamura, Y.; Lima, L.H.; Kurosaka, D.; Spaide, R.F. Choroidal Thickness and Visual Acuity in Highly Myopic Eyes. *Retina* **2012**, *32*, 1229–1236. [CrossRef] [PubMed]
15. Monaco, A.; Cattaneo, R.; Spadaro, A.; Giannoni, M.; Di Martino, S.; Gatto, R. Visual Input Effect on EMG Activity of Masticatory and Postural Muscles in Healthy and in Myopic Children. *Eur. J. Paediatr Dent.* **2006**, *7*, 18–22. [PubMed]
16. Zieliński, G.; Matysik-Woźniak, A.; Rapa, M.; Baszczowski, M.; Ginszt, M.; Zawadka, M.; Szkutnik, J.; Rejdak, R.; Gawda, P. The Influence of Visual Input on Electromyographic Patterns of Masticatory and Cervical Spine Muscles in Subjects with Myopia. *J. Clin. Med.* **2021**, *10*, 5376. [CrossRef]
17. Zieliński, G.; Filipiak, Z.; Ginszt, M.; Matysik-Woźniak, A.; Rejdak, R.; Gawda, P. The Organ of Vision and the Stomatognathic System—Review of Association Studies and Evidence-Based Discussion. *Brain Sci.* **2022**, *12*, 14. [CrossRef]
18. Fiorucci, E.; Bucci, G.; Cattaneo, R.; Monaco, A. The Measurement of Surface Electromyographic Signal in Rest Position for the Correct Prescription of Eyeglasses. *IEEE Trans. Instrum. Meas.* **2012**, *61*, 419–428. [CrossRef]
19. Czepita, M.; Czepita, D.; Safranow, K. Role of Gender in the Prevalence of Myopia among Polish Schoolchildren. *J. Ophthalmol.* **2019**, *2019*, 9748576. [CrossRef]
20. Czepita, D.; Mojsa, A.; Ustianowska, M.; Czepita, M.; Lachowicz, E. Role of Gender in the Occurrence of Refractive Errors. *Ann. Acad. Med. Stetin* **2007**, *53*, 5–7.
21. Jones-Jordan, L.A.; Sinnott, L.T.; Chu, R.H.; Cotter, S.A.; Kleinstein, R.N.; Manny, R.E.; Mutti, D.O.; Twelker, J.D.; Zadnik, K. CLEERE Study Group Myopia Progression as a Function of Sex, Age, and Ethnicity. *Investig. Ophthalmol. Vis. Sci.* **2021**, *62*, 36. [CrossRef]
22. Bueno, C.H.; Pereira, D.D.; Pattussi, M.P.; Grossi, P.K.; Grossi, M.L. Gender Differences in Temporomandibular Disorders in Adult Populational Studies: A Systematic Review and Meta-Analysis. *J. Oral Rehabil* **2018**, *45*, 720–729. [CrossRef] [PubMed]
23. Bagis, B.; Ayaz, E.A.; Turgut, S.; Durkan, R.; Özcan, M. Gender Difference in Prevalence of Signs and Symptoms of Temporomandibular Joint Disorders: A Retrospective Study on 243 Consecutive Patients. *Int. J. Med. Sci.* **2012**, *9*, 539–544. [CrossRef] [PubMed]
24. Satiroğlu, F.; Arun, T.; Işik, F. Comparative Data on Facial Morphology and Muscle Thickness Using Ultrasonography. *Eur. J. Orthod.* **2005**, *27*, 562–567. [CrossRef]
25. Chang, P.-H.; Chen, Y.-J.; Chang, K.-V.; Wu, W.-T.; Özçakar, L. Ultrasound Measurements of Superficial and Deep Masticatory Muscles in Various Postures: Reliability and Influencers. *Sci. Rep.* **2020**, *10*, 14357. [CrossRef]
26. Macrae, P.R.; Jones, R.D.; Myall, D.J.; Melzer, T.R.; Huckabee, M.-L. Cross-Sectional Area of the Anterior Belly of the Digastric Muscle: Comparison of MRI and Ultrasound Measures. *Dysphagia* **2013**, *28*, 375–380. [CrossRef]
27. Strini, P.J.S.A.; Strini, P.J.S.A.; de Souza Barbosa, T.; Gavião, M.B.D. Assessment of Thickness and Function of Masticatory and Cervical Muscles in Adults with and without Temporomandibular Disorders. *Arch. Oral Biol.* **2013**, *58*, 1100–1108. [CrossRef]
28. Ginszt, M.; Zieliński, G. Novel Functional Indices of Masticatory Muscle Activity. *J. Clin. Med.* **2021**, *10*, 1440. [CrossRef]
29. Hermens, H.J.; Freriks, B.; Disselhorst-Klug, C.; Rau, G. Development of Recommendations for SEMG Sensors and Sensor Placement Procedures. *J. Electromyogr. Kinesiol.* **2000**, *10*, 361–374. [CrossRef]
30. Shi, Q.; Wang, G.-Y.; Cheng, Y.-H.; Pei, C. Comparison of IOL-Master 700 and IOL-Master 500 Biometers in Ocular Biological Parameters of Adolescents. *Int. J. Ophthalmol.* **2021**, *14*, 1013–1017. [CrossRef]
31. Buckhurst, P.J.; Wolffsohn, J.S.; Shah, S.; Naroo, S.A.; Davies, L.N.; Berrow, E.J. A New Optical Low Coherence Reflectometry Device for Ocular Biometry in Cataract Patients. *Br. J. Ophthalmol.* **2009**, *93*, 949–953. [CrossRef]
32. Koman-Wierdak, E.; Róg, J.; Brzozowska, A.; Toro, M.D.; Bonfiglio, V.; Załuska-Ogryzek, K.; Karakuła-Juchnowicz, H.; Rejdak, R.; Nowomiejska, K. Analysis of the Peripapillary and Macular Regions Using OCT Angiography in Patients with Schizophrenia and Bipolar Disorder. *J. Clin. Med.* **2021**, *10*, 4131. [CrossRef] [PubMed]
33. Gella, L.; Raman, R.; Sharma, T. Macular Thickness Measurements Using Copernicus Spectral Domain Optical Coherence Tomography. *Saudi J. Ophthalmol.* **2015**, *29*, 121–125. [CrossRef] [PubMed]
34. Bonfiglio, V.; Ortisi, E.; Scollo, D.; Reibaldi, M.; Russo, A.; Pizzo, A.; Faro, G.; Macchi, I.; Fallico, M.; Toro, M.D.; et al. Vascular Changes after Vitrectomy for Rhegmatogenous Retinal Detachment: Optical Coherence Tomography Angiography Study. *Acta Ophthalmol.* **2019**, *98*, e563–e569. [CrossRef] [PubMed]
35. McKee, E.C.; Ely, A.L.; Duncan, J.E.; Dosunmu, E.O.; Freedman, S.F. A Comparison of Icare PRO and Tono-Pen XL Tonometers in Anesthetized Children. *J. AAPOS* **2015**, *19*, 332–337. [CrossRef]
36. Beneyto, P.; Barajas, M.A.; Garcia-de-Blas, F.; del Cura, I.; Sanz, T.; Vello, R.; Salvador, C. Predictive Value of Tonometry with Tono-Pen® XL in Primary Care. *Br. J. Gen. Pr.* **2007**, *57*, 653–654.

37. Cohen, J. *Statistical Power Analysis for the Behavioral Sciences*; Elsevier Science: Burlington, VT, USA, 2013; ISBN 978-1-4832-7648-9.
38. Fritz, C.O.; Morris, P.E.; Richler, J.J. Effect Size Estimates: Current Use, Calculations, and Interpretation. *J. Exp. Psychol Gen.* **2012**, *141*, 2–18. [CrossRef] [PubMed]
39. Lakens, D. Calculating and Reporting Effect Sizes to Facilitate Cumulative Science: A Practical Primer for *t*-Tests and ANOVAs. *Front. Psychol.* **2013**, *4*, 863. [CrossRef]
40. Gordon, R.A.; Donzis, P.B. Refractive Development of the Human Eye. *Arch. Ophthalmol.* **1985**, *103*, 785–789. [CrossRef]
41. Rozema, J.J.; Ní Dhubhghaill, S. Age-Related Axial Length Changes in Adults: A Review. *Ophthalmic Physiol. Opt.* **2020**, *40*, 710–717. [CrossRef]
42. Meng, W.; Butterworth, J.; Malecaze, F.; Calvas, P. Axial Length of Myopia: A Review of Current Research. *OPH* **2011**, *225*, 127–134. [CrossRef]
43. Zereid, F.M.; Osuagwu, U.L. Myopia and Regional Variations in Retinal Thickness in Healthy Eyes. *J. Ophthalmic Vis. Res.* **2020**, *15*, 178–186. [CrossRef] [PubMed]
44. Kong, M.; Choi, D.Y.; Han, G.; Song, Y.-M.; Park, S.Y.; Sung, J.; Hwang, S.; Ham, D.-I. Measurable Range of Subfoveal Choroidal Thickness With Conventional Spectral Domain Optical Coherence Tomography. *Transl. Vis. Sci. Technol.* **2018**, *7*, 16. [CrossRef] [PubMed]
45. Hoseini-Yazdi, H.; Vincent, S.J.; Collins, M.J.; Read, S.A.; Alonso-Caneiro, D. Wide-Field Choroidal Thickness in Myopes and Emmetropes. *Sci. Rep.* **2019**, *9*, 3474. [CrossRef] [PubMed]
46. Gupta, P.; Saw, S.-M.; Cheung, C.Y.; Girard, M.J.A.; Mari, J.M.; Bhargava, M.; Tan, C.; Tan, M.; Yang, A.; Tey, F.; et al. Choroidal Thickness and High Myopia: A Case–Control Study of Young Chinese Men in Singapore. *Acta Ophthalmol.* **2015**, *93*, e585–e592. [CrossRef] [PubMed]
47. Stecco, C.; Hammer, W.I. *Functional Atlas of the Human Fascial System*; Elsevier Ltd.: Edinburgh, UK, 2015; ISBN 978-0-7020-4430-4.
48. Sterniak, M. Terapia Krótkowzrocznosci Nabytej u Dzieci Metodą Manipulacji Powięzi—Studium Przypadku. *Fizjoterapia I Rechabilitacja* **2017**, *88*, 6–11.
49. AlShareef, S.; Newton, B.W. Accessory Nerve Injury. In *StatPearls*; StatPearls Publishing: Treasure Island, FL, USA, 2022.
50. Whitney, Z.B.; Jain, M.; Zito, P.M. Anatomy, Skin, Superficial Musculoaponeurotic System (SMAS) Fascia. In *StatPearls*; StatPearls Publishing: Treasure Island, FL, USA, 2022.
51. Davidge, K.M.; van Furth, W.R.; Agur, A.; Cusimano, M. Naming the Soft Tissue Layers of the Temporoparietal Region: Unifying Anatomic Terminology across Surgical Disciplines. *Neurosurgery* **2010**, *67*, ons120–ons129. [CrossRef]
52. Marasini, S.; Khadka, J.; Sthapit, P.R.K.; Sharma, R.; Nepal, B.P. Ocular Morbidity on Headache Ruled out of Systemic Causes—A Prevalence Study Carried out at a Community Based Hospital in Nepal. *J. Optom.* **2012**, *5*, 68–74. [CrossRef]
53. Bhatia, R.; Dureja, G.P.; Tripathi, M.; Bhattacharjee, M.; Bijlani, R.L.; Mathur, R. Role of Temporalis Muscle over Activity in Chronic Tension Type Headache: Effect of Yoga Based Management. *Indian J. Physiol. Pharm.* **2007**, *51*, 333–344.
54. Bendtsen, L.; Fernández-de-la-Peñas, C. The Role of Muscles in Tension-Type Headache. *Curr. Pain Headache Rep.* **2011**, *15*, 451–458. [CrossRef]
55. Zorena, K.; Gładysiak, A.; Ślęzak, D. Early Intervention and Nonpharmacological Therapy of Myopia in Young Adults. *J. Ophthalmol.* **2018**, *2018*, 4680603. [CrossRef]
56. Strini, P.J.S.A.; Strini, P.H.S.A.; Barbosa, T.S.; Gavião, M.B.D. Effects of Head Posture on Cervical Muscle Thickness and Activity in Young Adults With and Without Temporomandibular Disorders. *J. Musculoskelet. Pain* **2014**, *22*, 89–98. [CrossRef]
57. Grgić, V. Cervicogenic proprioceptive vertigo: Etiopathogenesis, clinical manifestations, diagnosis and therapy with special emphasis on manual therapy. *Lijec Vjesn* **2006**, *128*, 288–295. [PubMed]
58. Yahia, A.; Ghroubi, S.; Jribi, S.; Mâlla, J.; Baklouti, S.; Ghorbel, A.; Elleuch, M.H. Chronic Neck Pain and Vertigo: Is a True Balance Disorder Present? *Ann. Phys. Rehabil. Med.* **2009**, *52*, 556–567. [CrossRef] [PubMed]
59. Luan, H.; Gdowski, M.J.; Newlands, S.D.; Gdowski, G.T. Convergence of Vestibular and Neck Proprioceptive Sensory Signals in the Cerebellar Interpositus. *J. Neurosci.* **2013**, *33*, 1198–1210. [CrossRef]
60. Shankar Kikkeri, N.; Nagalli, S. Trigeminal Neuralgia. In *StatPearls*; StatPearls Publishing: Treasure Island, FL, USA, 2021.
61. Go, J.L.; Kim, P.E.; Zee, C.S. The Trigeminal Nerve. *Semin. Ultrasound CT MR* **2001**, *22*, 502–520. [CrossRef]
62. Zieliński, G.; Byś, A.; Szkutnik, J.; Majcher, P.; Ginszt, M. Electromyographic Patterns of Masticatory Muscles in Relation to Active Myofascial Trigger Points of the Upper Trapezius and Temporomandibular Disorders. *Diagnostics* **2021**, *11*, 580. [CrossRef]
63. Christensen, L.V.; Radue, J.T. Lateral Preference in Mastication: An Electromyographic Study. *J. Oral Rehabil.* **1985**, *12*, 429–434. [CrossRef]
64. Frayne, E.; Coulson, S.; Adams, R.; Croxson, G.; Waddington, G. Laterality of Proprioception in the Orofacial Muscles and Temporomandibular Joint. *Neurosci. Lett.* **2016**, *635*, 111–116. [CrossRef]
65. Pan, C.W.; Klein, B.E.; Cotch, M.F.; Shrager, S.; Klein, R.; Folsom, A.; Kronmal, R.; Shea, S.J.; Burke, G.L.; Saw, S.M.; et al. Racial Variations in the Prevalence of Refractive Errors in the United States: The Multi-Ethnic Study of Atherosclerosis. *Am. J. Ophthalmol.* **2013**, *155*, 1129–1138.e1. [CrossRef]
66. Luong, T.Q.; Shu, Y.-H.; Modjtahedi, B.S.; Fong, D.S.; Choudry, N.; Tanaka, Y.; Nau, C.L. Racial and Ethnic Differences in Myopia Progression in a Large, Diverse Cohort of Pediatric Patients. *Investig. Ophthalmol. Vis. Sci.* **2020**, *61*, 20. [CrossRef]

Journal of *Personalized Medicine*

MDPI

Review

Photoactivated Chromophore Corneal Collagen Cross-Linking for Infectious Keratitis (PACK-CXL)—A Comprehensive Review of Diagnostic and Prognostic Factors Involved in Therapeutic Indications and Contraindications

Ileana Ramona Barac [1,2], Andrada-Raluca Artamonov [2], George Baltă [1,2], Valentin Dinu [1,2,*], Claudia Mehedințu [3], Anca Bobircă [4], Florian Baltă [1,2] and Diana Andreea Barac [5]

1 Department of Ophthalmology/ENT, Faculty of Medicine, 'Carol Davila' University of Medicine and Pharmacy, 050747 Bucharest, Romania
2 Bucharest Emergency Eye Hospital, 030167 Bucharest, Romania
3 Department of Obstetrics and Gynecology, Faculty of Medicine, 'Carol Davila' University of Medicine and Pharmacy, 050747 Bucharest, Romania
4 Department of Rheumatology and Internal Medicine, Faculty of Medicine, 'Carol Davila' University of Medicine and Pharmacy, 050747 Bucharest, Romania
5 Faculty of Medicine, 'Carol Davila' University of Medicine and Pharmacy, 050747 Bucharest, Romania
* Correspondence: valentin.dinu@umfcd.ro; Tel.: +40-726195486

Abstract: Infectious keratitis is a severe infection of the eye, which requires urgent care in order to prevent permanent complications. Typical cases are usually diagnosed clinically, whereas severe cases also require additional tools, such as direct microscopy, corneal cultures, molecular techniques, or ophthalmic imaging. The initial treatment is empirical, based on the suspected etiology, and is later adjusted as needed. It ranges from topical administration of active substances to oral drugs, or to complex surgeries in advanced situations. A novel alternative is represented by Photoactivated Chromophore Corneal Collagen Cross-Linking (PACK-CXL), which is widely known as a minimally invasive therapy for corneal degenerations. The purpose of this review is to identify the main diagnostic and prognostic factors which further outline the indications and contraindications of PACK-CXL in infectious keratitis. Given the predominantly positive outcomes in the medical literature, we ponder whether this is a promising treatment modality, which should be further evaluated in a systematic, evidence-based manner in order to develop a clear treatment protocol for successful future results, especially in carefully selected cases.

Keywords: infectious keratitis; corneal ulcer; Photoactivated Chromophore Corneal Collagen Cross-Linking; PACK-CXL

Citation: Barac, I.R.; Artamonov, A.-R.; Baltă, G.; Dinu, V.; Mehedințu, C.; Bobircă, A.; Baltă, F.; Barac, D.A. Photoactivated Chromophore Corneal Collagen Cross-Linking for Infectious Keratitis (PACK-CXL)—A Comprehensive Review of Diagnostic and Prognostic Factors Involved in Therapeutic Indications and Contraindications. *J. Pers. Med.* **2022**, *12*, 1907. https://doi.org/10.3390/jpm12111907

Academic Editor: Chieh-Chih Tsai

Received: 13 October 2022
Accepted: 14 November 2022
Published: 16 November 2022

Publisher's Note: MDPI stays neutral with regard to jurisdictional claims in published maps and institutional affiliations.

1. Introduction

Infectious keratitis is defined as a pathological process in the cornea, caused by the presence of one or more pathogenic microorganisms. It is a medical emergency and if left untreated, it can lead to many complications, including corneal thinning, scarring, perforation, endophthalmitis, loss of sight, or even loss of the eye [1,2]. Unfavorable prognosis is suggested by an ulcer involving the visual axis, the presence of a large infiltrate, as well as low visual acuity at initial work-up, especially in the elderly [3]. The outcome depends on establishing a prompt, correct diagnosis, and then on choosing the most suitable intervention according to the causative agent [4].

Etiology can be either microbial (with bacteria, fungi, protozoa) or viral, and is pivotal to deciding the right treatment plan [5]. Initial therapy is empiric, but absolutely mandatory, because exacerbation is imminent, and diagnostic confirmation usually takes at least 48 h—wasted time being linked to a worse prognosis [1]. Depending on the causal agent, medical management includes topical administration of appropriate anti-infective drugs, cycloplegics, as well as subconjunctival or intrastromal injections, and oral therapy in selected cases. Surgical options consist of corneal biopsy, conjunctival flap, amniotic membrane transplantation, and therapeutic keratoplasty, and are reserved for advanced situations [1,6].

A viable therapeutic alternative is represented by Photoactivated Chromophore Corneal Collagen Cross-Linking (PACK-CXL), a procedure based on the properties of ultraviolet-A (UV-A) light radiations to photoactivate riboflavin (vitamin B2), leading to both stiffening of the corneal stroma and inactivation of pathogens [2,7].

2. Materials and Methods

The purpose of this paper is to comprehensively review the medical literature on the specific diagnostic and prognostic factors which dictate the indications and contraindications of PACK-CXL in infectious keratitis, while tracing the history, development, and outcomes of this procedure over the last 15 years—from its first implementation until 27 July 2022.

Although corneal cross-linking represents an already recognized tool in ophthalmology, it has not been widely established in general clinical practice yet, and it is still under evaluation by the scientific community in the treatment of corneal infections. Therefore, we believe such reviews are necessary until guidelines are thoroughly defined and applied.

In this regard, our search was conducted on PubMed and Google Scholar, on 27th of July 2022, using the following keyword: '((corneal OR collagen) cross-linking OR photoactivated chromophore OR CXL OR PACK-CXL) AND (keratitis OR corneal ulcer)'. We used both MeSH (PubMed) and free-text vocabulary. In order to be as comprehensive as possible, we included case reports, case series, clinical and epidemiological studies, reviews, and meta-analyses dating as early as 2008, with significant impact, clear methodology, and direct relevance to the subject, as appreciated by the authors of this paper (Figure 1). We considered all articles available in English concerned with the assessment of PACK-CXL in the context of infectious keratitis. We excluded articles about non-infectious keratitis; keratitis caused by CXL; non-infectious corneal melting or ulcer; general reviews about keratitis with brief mentioning of CXL; general reviews about cross-linking or the use cross-linking procedure in the context of various corneal ectasias and other keratopathies; combined CXL and surgical procedures; and cases pertaining to veterinary medicine. Of the 337 papers screened, 55 have been included (Table A1).

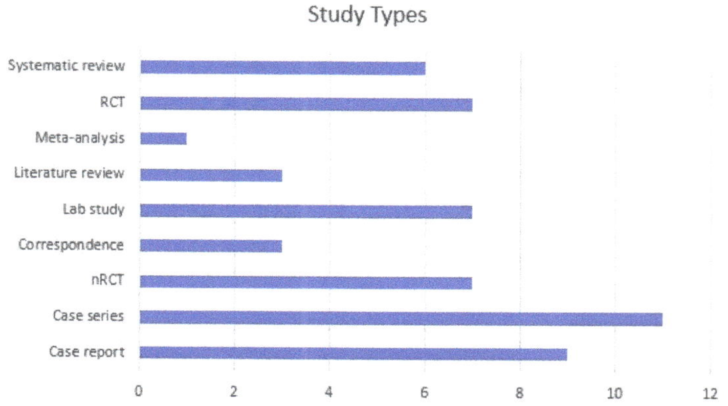

Figure 1. Bar graph showing distribution of study types selected for review. (RCT: randomized controlled trial, nRCT: non-randomized clinical trial).

3. Results

3.1. The Challenge of Accuracy

The precise recognition of the pathogen is a process in itself and comprises multiple steps. Clinical examination and microbiological techniques play a central role, but newer investigation methods are emerging and could further complete the diagnostic picture [1]. However, it is important to note the fact that the standard approach is sometimes flawed, considering that observable features do not always follow classical descriptions, that they might be altered or combined in the presence of multiple microorganisms, that cultures take a certain time to grow, and that microbiological detection is not guaranteed, given the high percent of false negatives. Fortunately, the more severe cases, characterized by larger lesions with higher microorganism densities, lead to greater positivity rates in both smears and cultures [4]. Even so, findings in the medical literature indicate that more than 20% of the cases are polymicrobial and more than 50% do not have a positive culture identification of an infectious agent [6].

Upon presentation, previous ocular history might be suggestive of the cause. Contact lens wear is frequently linked to a bacterial (Pseudomonas aeruginosa), fungal, or parasitic infection (Acanthamoeba), and corneal lacerations with vegetal matter (i.e., branches, leaves) may be complicated by a fungal infection [4]. A recent LASIK intervention points towards certain atypical pathogens, including fungi (e.g., Exophiala dermatitidis), non-tuberculous mycobacteria (e.g., Mycobacterium chelonae), and other bacteria (Nocardia, but also methicillin-resistant Staphylococcus aureus or Streptococcus spp.) [6]. However, symptoms are variable and rather non-specific, including decreased vision, discharge, photophobia, tearing, and foreign body sensation with inconsistent levels of pain.

Slit-lamp examination usually reveals infiltrate(s), haze, edema or ulceration, with particular characteristics depending on the cause. Typically, the infiltrate is localized, round-shaped and well-defined in Gram-positive infections, but it is suppurative and rapidly progressing in Gram-negative cases [1]. If it is determined by filamentous fungi, it may appear as an elevated and dry lesion or as an endothelial plaque, but if it is due to Candida, it may be a slowly evolving stromal lesion. Notably, herpetic ulcers are dendritic or geographic, whereas Acanthamoeba induces irregularities in the epithelium [6]. The stromal infiltrates caused by Gram-positive bacteria and Candida are distinct and white-grey in color, whereas those caused by Gram-negative bacteria and Acanthamoeba usually form an immune ring, and those by filamentous fungi have feathery borders and multiple satellite lesions (the later also suggesting the appearance of Acanthamoeba infiltrates) [1]. Gram-positive keratitis presents with minimal haze, in contrast to Gram-negative cases, whereas filamentous fungus infections develop endothelial plaques. Moreover, there are individual characteristics pertaining to some species. Nocardia determines white, numerous infiltrates, arranged in a wreath, with accompanying fine filaments towards healthy cornea. On the other hand, non-tuberculous mycobacteria keratitis has been described as having a 'cracked windshield' appearance. Microsporidium can be mistaken for atypical adenoviral keratoconjunctivitis, due to its punctate epithelial lesions and subepithelial scarring [6]. Viral keratitides are the most heterogeneous. The herpes simplex virus itself determines either epithelial, stromal, endothelial or kerato-uveitic presentations, whereas the varicella-zoster virus typically manifests as nummular keratitis [1].

Despite their detailed descriptions, these findings are not consistent and can be unreliable in clinical settings, as mentioned above. It has been shown that cornea-trained specialists correctly distinguish bacterial keratitis from fungal keratitis in 66% of the cases [8]. This is why the objective identification of the etiology through microbiology techniques is oftentimes employed, and primarily consists of staining and microscopy examination, cultures, and sensitivity testing [4]. Theoretically, this should represent the protocol in all suspected infections [5], but most often, clinicians, through no fault of their own, resort to treating the majority of cases empirically, based on experience. This is in part due to the fact that the actual usefulness of additional diagnostic steps remains a controversial topic, taking into consideration that they are more expensive and time-consuming, and that broad-spectrum antibiotics often lead to good outcomes, especially in small, superficial, peripheral ulcers of non-traumatic origin [9]. Certain guidelines state that such investigations are only needed in advanced cases of keratitis, which involve the visual axis, consist of large or multiple lesions, display atypical characteristics, appear in the context of recent corneal surgery, or do not respond to broad-spectrum therapy [1], and several papers have underlined that they are actually employed in only 5% to 15% of the cases [4].

In order to identify the pathogen, the first part is represented by corneal scrapings under topical anesthesia, ideally after removing mucus and debris, primarily targeting the lesion's base and active borders [4]. Then, samples are smeared onto slides stained with various substances and analyzed using direct microscopy. The most widely used stains are Gram—for bacteria, Giemsa or potassium hydroxide (KOH)—for fungi, and Calcofluor White Staining (CFW) or lactophenol for Acanthamoeba [9]. Gram is common and well-standardized in classifying cocci and bacilli, but errors due to both logistics and interpretation generate vast variability between laboratories, with reported sensitivity ranging from 30% to 100%. In addition, there can remain unstained bacteria, such as Mycobacterium spp. or Nocardia spp. Therefore, Ziehl-Nielsen (ZN) or Kinyoun (modified ZN) could be necessary in selected cases [4]. On the other hand, yeasts, including Candida, might be missed under Giemsa, or mistaken for artifacts [9].

The second use of corneal scrapings material is for cultures, which is the definitive gold standard in current medical practice and should always include appropriate media for both bacteria and fungi [4,9]. Apart from corneal scrapings, materials such as contact lenses, their cases and cleaning solutions, or loose sutures on the eye surface, which have been in direct contact with infected tissues, can also be cultured [4]. The positivity rate varies greatly, similar to direct microscopy, and is influenced by multiple factors, such as technical difficulties, delays, low pathogen load, clinical severity, or recent use of anti-infective agents or of topical steroids [4–6]. In addition, pathogens have to be differentiated from commensal microorganisms, using a multitude of criteria. Logistics include a variety of well-defined protocols which can be employed, and they include direct and indirect inoculations on solid agars or liquids [4]. For common bacteria, usual choices encompass blood agar, chocolate agar, or brain–heart infusion at 37°, whereas Mycobacterium grows on Lowenstein–Jensen and Middlebrook media. Fungi can develop in similar conditions if antibacterials are added, or can grow on Sabouraud dextrose agar. Acanthamoeba needs non-nutrient agar, with an E. coli overlay [1,6]. Then, these media are incubated in strict conditions for 1–21 days, and are re-examined daily for noticeable changes [9]. Typical bacteria grow relatively fast, but Nocardia, Mycobacterium, fungi, and Acanthamoeba require significantly more time, which can sometimes be unpredictably long [4].

Another point of discussion is concerned with antibiotic sensitivity and resistance, as reactionary patterns to the readily available ophthalmic drugs are changing permanently [9]. However, laboratory methods determine systemic concentrations, rather than topical ones, which can be confusing, as they do not take into consideration the direct route of administration, the frequency of instillations, or the fortified forms used in ophthalmology. Therefore, it is advisable that the treatment course should be planned by integrating both microbiological data and timely assessments of the patients' clinical evolution [4].

All the aforementioned information is directed towards the main diagnostic proceedings in microbial infections, i.e., bacterial, fungal or parasite cases. As far as viral keratitis is involved, diagnosis is usually clinical, given the particular appearance at the slit-lamp and the additional presence of herpetic vesicles [1]. The arsenal for objective confirmation is fundamentally different from that of other types of pathogens, and comprises of cultures on Vero cell lines, direct antibody identification using fluorescence, and Polymerase Chain Reaction (PCR) from active lesions or tears samples [5,6].

Consequently, the most common pathogens involved in keratitides are the following:

- Gram-positive bacteria: coagulase-negative staphylococci, Staphylococcus aureus, Streptococcus pneumoniae;

- Gram-negative bacteria: Pseudomonas aeruginosa, Enterobacteriaceae, Moraxella, Haemophilus, Neisseria gonorrhoeae;
- acid-fast bacteria: Mycobacterium, Nocardia;
- filamentous fungi: Fusarium, Aspergillus, Curvularia, Alternaria;
- yeasts: Candida albicans, other Candida spp., Cryptococcus;
- fungi-like: Microsporidium;
- parasites: Acanthamoeba;
- viruses: HSV 1, CMV. VZV, Adenovirus [1,4,6].

In progressing non-responsive cases with previous sterile cultures, additional diagnostic techniques are employed. Corneal biopsy might be performed in order to gain deeper access to corneal infiltrates and to facilitate histopathology analyses, apart from direct microscopy and culturing, which clarifies a potential fungal or Acanthamoeba etiology. However, clinicians should be wary of perforation risks, especially with thinning, melting, or necrotic tissue [4]. Similarly, impression cytology could be used for the same purposes, as it is especially helpful for fungi or parasite cases [6], whereas Transmission Electron Microscopy (TEM) represents the gold standard for confirming the presence of Microsporidium spores [1].

Moreover, PCR can also be used in all microbial infections, with faster results and high sensitivity, including pathogens with slow and difficult growth on cultures. Its disadvantages include high equipment and training costs, the need to target microorganisms specifically by picking the right primers, and lower specificity, as it does not discriminate active from dead pathogens, or from background flora [1]. An alternative molecular diagnosis method is represented by mass spectrometry, which might play an important role in identifying rare species [4]. Over time, more sophisticated techniques have been developed, such as metagenomics next-generation sequencing (NGS), which is based on nucleic acid amplification and shares similar working principles with PCR but is currently more suitable to research than to clinical use [1].

Novel ophthalmic non-invasive imaging tools have been proposed within diagnostic frameworks. For this scope, in vivo confocal microscopy (IVCM) has been used to generate real-time images with a resolution of 1 μm, which is sensitive enough in order to identify fungal filaments or parasitic cysts—but not smaller pathogens—anywhere in the depth of the cornea. However, it is expensive, heavily operator-dependent, and it generates potential errors due to artifacts [1,9]. Anterior Segment Optical Coherence Tomography (AS-OCT) is a more accessible method, with excellent anatomical and pathological descriptions, as well as precise quantitative data. It can be used in conjunction with biomicroscopy, when deep ulcers or large infiltrates cannot be assessed properly and obscure deeper ocular tissues. It offers valuable information on corneal thickness, which has been shown to fluctuate proportionately with inflammation severity, and can therefore be useful in monitoring progress, but cannot identify the responsible pathogen [4,10].

Lastly, Artificial Intelligence shows immense promise in analyzing images, with accurate interpretations through the means of well-trained pattern recognition algorithms. Thus, it can differentiate active lesions from corneal scars, or typical bacterial and fungal ulcers. For the moment, the most important real-life utility could be the rapid screening, diagnosis, and appropriate recommendation-making based on external photos of the eye, in the frame of a telemedical service for communities with low access to an ophthalmologist [1].

3.2. The Challenge of Choice

It is important to underline the fact that successful treatment of infectious keratitis is linked to the accurate identification of the responsible pathogen [4].

Therefore, typical treatment in bacterial cases consists of broad-spectrum antibiotics with topical administration, which should be initiated empirically as soon as possible—eventually, after collecting appropriate samples for laboratory analyses, if indicated. Gold standard schemes include either fluoroquinolone monotherapy, or fortified combinations of cephalosporins and aminoglycosides [1]. The subconjunctival route might be preferred in specific cases, when the risk of perforation is high or adherence issues might occur, whereas systemic therapy is necessary if the infection keeps spreading towards adjacent tissues. The two main medical alternatives have been shown to be similar in efficacy, and the choice depends on clinical judgment, experience of the physician, and drug availability [11]. However, growing resistance towards antibacterial substances represents a major public health issue, posing a challenge in specific cases, especially those of staphylococcal or Pseudomonas origin. The impact of this phenomenon is currently not as defined in ophthalmology as it is in systemic infections, but prudence is advisable in order to prevent a future decline in susceptibility to common antibiotics, especially in typical cases that are relatively straightforward to treat in today's climate [5].

On the other hand, fungal keratitis is medically cured with antimycotic therapy, but generally has a worse prognosis compared to bacterial ulcers, due to lower drug penetration levels, as well as diagnosis difficulties [1]. Topical administration includes several alternatives, such as 5% natamycin or 1% itraconazole for filamentous fungi, and 0.15% amphotericin B or fluconazole for yeasts [6]. Intracameral or intrastromal injections are useful in extended or non-responsive infections [1], and various oral triazoles are also available [6].

An even more complicated situation is represented by Acanthamoeba infections, as timely diagnosis and aggressive treatments are fundamental to satisfactory clinical evolution. Combinations of medicine must be used, because an agent capable of eliminating both trophozoites and cysts does not exist [1]. However, dibromopropamidine, hexamidine, chlorhexidine, and polyhexamethyl-biguanide are acceptable topical options [6].

As far as viral keratitis is concerned, treatment remains controversial to some degree and depends on the site of the infection. If only the corneal epithelium is affected, topical therapy with acyclovir should suffice. However, if there is stromal involvement, topical steroids are mandatory, with careful monitoring of local complications, especially a rise in IOP [1,6].

Regardless of the cause, adjunctive therapy plays an additional role in catalyzing the good clinical outcome of all cases. Regular debridement and saline instillations help removing necrotic tissue and secretions, thus decreasing local pathogen load and increasing treatment penetration [12]. Cycloplegics are prescribed in order to decrease synechiae development and pain when there is remarkable inflammation in the anterior chamber. On the other hand, the role of topical steroids in bacterial cases is unclear, the risk-benefit balance is not yet calibrated by definitive evidence, and its use should be judicious and dictated by case particularities, such as visual axis involvement, and an already good response to antibiotic therapy [11].

Even with prompt medical treatment, results often remain poor due to complications, and severe cases can lead to rapid deterioration, with no alternative other than surgical treatment. This includes amniotic membrane transplantation and penetrating keratoplasty [1,6]. However, operating on an infected eye is especially risky and prone to complications and failure, so that evisceration and enucleation are indicated in extreme cases where the visual potential has been lost [12].

3.3. *The Challenge of Novelty*

In recent years, an alternative therapy has been proposed for infectious keratitis, represented by Corneal Collagen Cross-Linking (CXL). It has already been used with good outcomes, as an adjuvant to antimicrobial therapy in patients with treatment-resistant corneal infections, bacterial-only [13–17], fungal-only [18–20], in mixed bacterial-fungal cases [2,21–25], or in Acanthamoeba infections [26–28]. Some investigators focused exclusively on refractory corneal ulcers and found that cross-linking therapy is beneficial [29–32], and that it can also markedly reduce healing time [33].

The effectiveness of CXL as a primary therapy has been shown both in animal models [34,35], and in clinical studies [36], but other authors argued that this approach has no advantage over the standard treatment [37], and a recent randomized clinical trial underlined that the clear benefit of CXL per primam therapy could not be yet proven, despite a lower complication rate in the cross-linking-only group [38].

The advantages of this minimally invasive approach are reinforced by a study focused on the one-year follow-up after CXL, which confirms the favorable prognosis long-term post-procedure [39].

CXL uses riboflavin (vitamin B2) drops, which act as a chromophore, photoactivated by UV-A radiations at 365–370 nm wavelength. Thus, free radicals are generated, promoting both an increase in collagen fiber diameter, as well as the creation of additional collagen-proteoglycan bonds in the corneal stroma. These new cross-links increase biomechanical stiffness, useful in the treatment of ectatic disorders of the cornea, markedly keratoconus, which represents the most widely-known scope of this minimally-invasive medical procedure [7,40,41].

In 2003, Wollensak, Sporl, and Seiler reported the preliminary successful results of CXL in stopping keratoconus progression, in a non-randomized five-year study, using a specific set of steps which would later be called "the Dresden protocol": under sterile conditions in the operating room, the central corneal epithelium is removed under local anesthesia; 0.1% riboflavin solution is applied every 5 min for 30 min; then, 370 nm UV-A irradiation begins, using a lamp, 1 cm away from the cornea, for 30 min, with an intensity of 3 mW/cm^2, translating to a total amount of energy of 5.4 J/cm^2; finally, a bandage contact lens is applied. Topical antibiotic drops are administered until reepithelialization is noted [42]. In accordance with the Bunsen–Roscoe Law of Reciprocity of Photochemistry, which asserts that the effects of UV-A radiation and the final dose of energy are directly correlated, regardless of the combinations between illumination time and intensity of light, as long as the total amount of energy remains the same [43], modifications to this standardized model have been proposed. In order to improve efficiency in this rather time-consuming technique, ophthalmologists employed UV-A rays of higher intensity, thereby shortening the procedure and consequently decreasing the chance of corneal dehydration, while increasing the comfort of both patient and doctor, with beneficial results without supplementary risks [44].

Furthermore, an additional role of riboflavin in combination with UV-A light includes pathogen inactivation, by inducing DNA damage in bacteria and viruses, to a degree which appears to make it more difficult to repair (compared to degradation suffered by host cells) [45], and also by increasing resistance against protein digesting enzymes, such as collagenase, trypsin, and pepsin (similar to the metalloproteinases involved in corneal ulcers) [46]. Clinically, this application has been demonstrated in transfusion medicine, by microbial decontamination of blood products [47,48]. Interestingly, an initial variant of the cross-linking procedure had actually been successfully employed by Schnitzler, Spoerl, and Seiler in 2000, in the treatment of non-infectious corneal melting, even before the publication of the classical Dresden protocol for CXL in keratoconus [42,49]. Furthermore, in 2008, Iseli et al. conducted the first study to assess the efficacy of CXL in treatment-resistant microbial keratitis, with promising results [50], and in 2012, Makdoumi et al. performed a pilot study in order to evaluate CXL as primary therapy in bacterial keratitis, with no prior antibiotic administration, with yet another favorable assessment [36].

Accelerated protocols have also been employed (either 9 mW/cm^2 for 10 min, 18 mW/cm^2 for 5 min, 36 mW/cm^2 for 2.5 min), all reaching the same positive conclusions towards a more efficient technique [51–54]. Further studies might bring other optimized techniques, such as the positive link between higher UV fluence and increased levels of microbial killing [28,55], as well as higher concentration of riboflavin [56].

Following all these advancements, at the 2013 meeting of the International Congress of Corneal Cross-Linking in Dublin, a clear difference has been established between the treatment technique in keratoconus (from then on, simply called Corneal Cross-Linking or CXL) and the treatment of infectious keratitis (from then on, called Photoactivated Chromophore for Infectious Keratitis or PACK-CXL) [7].

Additional indisputable advantages are represented by the fact that PACK-CXL does not further contribute to antibiotic resistance, as corneas are considered sterile after the procedure. Along with the financial benefits, there were reductions across three areas: cost of medication itself, number of follow-ups, and potential hospitalizations [25,52]. Another interesting finding is that, apart from the great clinical feedback, there is electrophysiological proof that cross-linking does not damage the retina and the optic nerve [57].

However, success rates might be unequal among pathogen types, with Gram-negative bacteria being the most susceptible and fungi, the least [58]. Antibiotic resistance does not seem to be correlated with photooxidative stress resistance [59]. One recent meta-analysis underlined rigorous evidence in case of bacterial infection, suggesting further inquiries are needed for those of fungal, parasite, or viral origin [60]. On the other hand, other reviews raised concerns about the deep clinical and methodological heterogeneity of the available literature, and actually refrained from drawing definitive conclusions [61,62].

Among potential complications of PACK-CXL, a meta-analysis noted corneal edema, loss of endothelial cells, and disease progression with decompensation, leading to perforation in rare cases. This might be influenced by the corneal depth of the infiltrates, as more than 250 microns increase both the risk of endothelial cell loss and resistance to the effect of PACK-CXL [63,64]. It is essential to take note of the fact that the procedure itself could facilitate infections, considering that the step of epithelial debridement removes the physical barrier of the cornea, thus exposing it to contagious agents [65]. Moreover, this procedure is cytotoxic to keratocytes, up to 300 microns, but especially in the first 100 microns, which absorb half of the total energy [66]. A major observation is the reactivation of previous herpes simplex keratitis following CXL, which makes it a contraindication [24,63], thus illustrating the importance of an accurate diagnosis.

Additional concerns have been expressed regarding the efficiency of PACK-CXL as primary treatment in fungal keratitis, as it might not lead to expected outcomes in deep stromal infections [58,67,68], by either not showing clinical advantages over conservative therapy [69,70], or even by leading to worse results when compared to medication [71,72]. However, these negative outcomes have been highly debated by scientists in the field [73,74]. Therefore, conclusions remain unclear.

Important differences have also been found among re-epithelization periods for different microorganisms. Whereas ulcers of bacterial origin can heal in as fast as 3 days, fungal and protozoa keratitides can take much longer (up to more than 100 days). Among bacteria, Mycobacterium is the most problematic, also needing more than 100 days for the corneal wound to close [63].

On the other hand, a recent randomized, prospective, phase 3 trial published in 2022 by Hafezi et al. underlined no difference in major complications between medically treated patients and the cross-linking group [25]. Given the complexities of these conflicting literature findings, more inquiries are needed in order to assess these problematic aspects, as already suggested by previous studies [66,75].

4. Discussion

It is known that infectious keratitis is a potentially dangerous condition, in which the precise diagnosis and accurate treatment are paramount. In addition, appropriate timing is of utmost importance in order to prevent eye-threatening complications. However, the gold standard therapy is well-established, and it renders good results in a majority of cases. Therefore, we raised the question if there is a place for PACK-CXL in everyday medical practice by identifying its possible indications and contraindications, from diagnostic and prognostic points of view.

As shown above, the medical literature confirmed the efficacy of PACK-CXL in a variety of case series and small studies conducted by multiple physicians throughout the world. Consequently, we attempted to identify a string of common particularities in a majority of these cases, regarding diagnosis difficulties (Table 1).

Table 1. Relevant causes for diagnostic failure.

Diagnostic Difficulties
Repeated sterile cultures, either caused by unsuitable previous use of anti-infective medicine, or by the supposed presence of fastidious or rare pathogens
Polymicrobial infections, in which some of the responsible agents are not identified
Lack of access to advanced diagnostic techniques, either caused by lack of funding, available technology, or trained specialists

In addition, we also underline a variety of prognostic factors, shared among many of the situations when cross-linking was used (Table 2).

Table 2. Clinical scenarios where CXL was employed.

Clinical Scenarios
Severe, advanced cases, with late presentation
Cases non-responsive to usual therapy, progressive despite correct medical treatment
Infections involving the visual axis
Cases with ominous signs of imminent complications

Moreover, by applying similar reasoning, we propose several scenarios when PACK-CXL could be indicated (Table 3).

Table 3. Possible PACK-CXL indications.

PACK-CXL Indications	Reasoning
Polymicrobial infections, even if not all of them have been identified	To reduce treatment costs, to improve adherence and, ultimately, to spare the patient from the exposure to multiple potent drugs and their possible adverse effects
Documented resistance to the available anti-infective agents, or remarkable shifts in local susceptibility patterns	To obviate potential future issues in the community
Corneal ulcers following trauma with significant contamination	To reduce microbial load as quickly as possible
Patients with severe keratitis and monocular vision	To reduce microbial load as quickly as possible
Allergies, sensitivity, or contraindications to the recommended medical therapy	To help preserve the ocular surface and to reduce the inflammatory response
History or suspicion of poor adherence	To reduce the need for long-term therapy
Vulnerable populations (pregnant women, elderly patients)	for whom potent systemic therapies or surgeries could be detrimental

However, contraindications of this procedure are not as clear, and neither is their absolute or relative nature. Some of them can be derived from the exclusion criteria of the initial study of safety of CXL in corneal dystrophies [76] (Table 4).

We must acknowledge the fact that this randomized clinical trial, which led to the FDA approval of Corneal Collagen Cross-Linking in the treatment of keratoconus, only excluded certain patient categories, considered at risk for complications or therapeutic failure. It has not explicitly demonstrated that these criteria also translate to contraindications. For instance, later studies extended the age range to 8-year-olds [43], and further input is needed in order to confirm each criterion and to define even more precise parameters, based on indisputable evidence.

Other contraindications of PACK-CXL in the context of corneal ulcers, as mentioned above in the review section, are keratitis of a viral cause that infiltrates deeper than 250 microns.

Table 4. CXL contraindications.

PACK-CXL Contraindications
Allergies, sensitivity, or contraindications to riboflavin, to local anesthetics, or to any other materials used during the procedure
Corneal thickness of less than 375 microns, before debridement of the epithelium
History or likelihood of delayed corneal wound healing
Significant corneal scarring or opacification
History of viral keratitis
Aphakia, or pseudophakia with a non-UV-blocking lens
Nystagmus, or any disorder which might interfere with a steady gaze
Pregnancy or nursing
Age under 12 years old

Considering the current medical climate, which has been built on evidence over the last decades and is dominated by protocols, a multitude of reservations about PACK-CXL occur, halting its large adoption in the treatment of microbial keratitis, at least for the moment. Some reasons for this reluctance include the lack of systematic research and the lack of official approval from international regulatory organisms. These two main arguments are doubled by occasional publication of papers with questionable research methodologies, heterogeneous results, lack of reproducibility, and surrounding controversy, which makes conclusions harder to draw. This uncertainty is especially relevant when the cross-linking procedure comes into discussion for the most difficult and advanced cases, where the treatment course has to be chosen carefully in order to prevent the imminent ocular damage.

It is still unclear if PACK-CXL should be used per primam, without recent history of medical treatment failure, or in relatively mild cases. Moreover, it has not yet been tested in certain populations, such as young children or pregnant patients, which might benefit from it (at least in comparison to surgery). However, it is important not to fall into the trap of picking a novel procedure for the sake of novelty only, especially until safety is assured. There are still many unanswered questions: Should this be an adjuvant to antimicrobial treatment, or be used per primam? What would the indications and contraindications be in relation to corneal thickness? What kind of protocol (standard, intermediate, fast) is best suited? What kind of riboflavin should be used to maximize results? How many times can the procedure be repeated? Under which conditions should patient from vulnerable populations receive this therapy?

The strengths of this paper consist of a thorough inquiry of available research information, culminating in the most cohesive list of potential indications and contraindications of PACK-CXL in infectious keratitis at the moment (as far as we managed to find through our review), based on critical thinking and appraisal; a well-defined keyword and search framework, facilitating a precise exploration of medical literature on the chosen subject, suitable for the academic purpose of this investigation; the large number of screened articles, which helped to trace the development of cross-linking back to its inception; and a comprehensive review of the diagnostic and prognostic factors of infectious keratitis, and an extensive description of CXL, including history, working mechanisms, various protocols, and results and weaknesses, which creates a greater perspective on the topic.

The limitations include the narrative nature of the review, which does not comply with PRISMA guidelines and is prone to subjectivity issues; the lack of in-depth mentions, analyses, and comparisons of methodology for the papers included in this review; and the absence of statistical confirmation for our descriptive observations and qualitative conclusions.

5. Conclusions

PACK-CXL represents a promising treatment for microbial keratitis, with multiple significant advantages—it is not influenced by the type of pathogen and its characteristics (including resistance to medication), it is minimally invasive, easy and safe to perform, and it leads to great clinical and imaging outcomes, with decreased costs and increased patient comfort. Over the following years, it could become an effective and well-established tool in battling difficult cases, minimizing or even removing the need for surgery, as well as preventing other serious sequelae, such as corneal perforation. It is believed to represent an effective, safe alternative to traditional medical treatment, yet more systematic research is needed in order to establish the exact indications and the specific protocols.

Author Contributions: Conceptualization, I.R.B. and A.-R.A.; methodology, A.-R.A.; software, G.B.; validation, V.D., I.R.B. and A.B.; formal analysis, V.D. and D.A.B.; investigation, I.R.B.; resources, F.B.; data curation, C.M.; writing—original draft preparation, I.R.B. and A.B.; writing—review and editing, I.R.B. and V.D.; visualization, D.A.B.; supervision, I.R.B.; and project administration, C.M. and F.B. All authors have read and agreed to the published version of the manuscript.

Funding: This research received no external funding.

Institutional Review Board Statement: Not applicable.

Informed Consent Statement: Not applicable.

Conflicts of Interest: The authors declare no conflict of interest.

Appendix A

Table A1. Complete list of papers included in the review.

Authors	Date Published	Journal
B. Knyazer et al.	April 2018	Cornea
D. Tabibian et al.	December 2014	J. Refract. Surg.
I. R. Barac et al.	March 2021	Exp. Ther. Med.
O. Richoz et al.	2014	J. Refract. Surg.
J. B. Tayapad et al.	July 2013	Curr. Opin. Ophthalmol.
P. Garg et al.	January 2017	Middle East Afr. J. Ophthalmol.
R. Sorkhabi et al.	February 2013	Int. Ophthalmol.
R. Shetty et al.	August 2014	Br. J. Ophthalmol.
B. I. Ramona et al.	January 2016	Rom. J. Ophthalmol.
D. G. Said et al.	2014	Ophthalmology
E. A. Awad et al.	2020	Int. J. Ophthalmol.
R. Deshmukh et al.	October 2019	Indian J. Ophthalmol.
R. Awad et al.	April 2022	Eur. J. Ophthalmol.
S. A. Davis et al.	June 2020	Cochrane Database Syst. Rev.
L. Papaioannou et al.	January 2016	Cornea
T. C. Y. Chan et al.	December 2015	Acta Ophthalmol.
E. Erdem et al.	June 2018	Mycopathologia
M. Zamani et al.	January 2015	J. Ophthalmic Vis. Res.
J. L. Alio et al.	2013	J. Ophthalmic Inflamm. Infect.
M. Uddaraju et al.	July 2015	Am. J. Ophthalmol.
Á. Arance-Gil et al.	2014	Cont. Lens Anterior Eye
M. O. Price and F. W. Price et al.	2016	Curr. Opin. Ophthalmol.
P. Rosetta et al.	December 2018	Case Rep. Ophthalmol. Med.
N. V. Prajna et al.	July 2021	Cornea
A. Yagci et al.	October 2016	Exp. Clin. Transplant.
N. V. Prajna et al.	February 2020	Ophthalmology
A. Abbouda et al.	April 2018	Semin. Ophthalmol.
O. Zloto et al.	August 2018	J. Refract. Surg.

Table A1. *Cont.*

Authors	Date Published	Journal
K. Tal et al.	July 2015	Cornea
T. M. Ferrari et al.	April 2009	Eur. J. Ophthalmol.
K. Bilgihan et al.	August 2016	Curr. Eye Res.
R. B. Vajpayee et al.	March 2015	Clin. Experiment. Ophthalmol.
A. Wei et al.	July 2019	Graefes Arch. Clin. Exp. Ophthalmol.
S. Kling et al.	August 2020	Cornea
M. Nateghi Pettersson et al.	September 2019	Am. J. Ophthalmol. case reports
P. Basaiawmoit et al.	2018	Cornea
F. Hafezi et al.	January 2022	Eye Vis. 2022 91
D. Tabibian et al.	January 2015	J. Ophthalmic Vis. Res.
E. A. Idrus et al.	November 2019	Acta Ophthalmol.
D. S. J. Ting et al.	October 2019	Ocul. Surf.
N. Kasetsuwan et al.	May 2016	Am. J. Ophthalmol.
A. Panda et al.	October 2012	Cornea
M. O. Price et al.	October 2012	J. Refract. Surg.
K. Makdoumi and A. Bäckman et al.	September 2016	Clin. Experiment. Ophthalmol.
D. S. J. Ting et al.	August 2020	Ophthalmology
D. Singhal et al.	January 2021	Ophthalmology
A. Skaat et al.	July 2013	Eur. J. Ophthalmol.
G. Galperin et al.	February 2012	Cornea
E. Chan et al.	September 2014	J. Cataract Refract. Surg.
S. H. Watson et al.	March 2022	Am. J. Ophthalmol. case reports
A. Saglík et al.	November 2013	Eye Contact Lens
S. Bamdad et al.	March 2015	Cornea
H. P. Iseli et al.	June 2008	Cornea
N. Al-Sabai et al.	2010	Bull. Soc. Belge Ophtalmol.
K. Makdoumi et al.	January 2012	Graefes Arch. Clin. Exp. Ophthalmol.

References

1. Cabrera-Aguas, M.; Khoo, P.; Watson, S.L. Infectious keratitis: A review. *Clin. Exp. Ophthalmol.* **2022**, *50*, 543–562. [CrossRef] [PubMed]
2. Papaioannou, L.; Miligkos, M.; Papathanassiou, M. Corneal Collagen Cross-Linking for Infectious Keratitis: A Systematic Review and Meta-Analysis. *Cornea* **2016**, *35*, 62–71. [CrossRef] [PubMed]
3. Ting, D.S.J.; Cairns, J.; Gopal, B.P.; Ho, C.S.; Krstic, L.; Elsahn, A.; Lister, M.; Said, D.G.; Dua, H.S. Risk Factors, Clinical Outcomes, and Prognostic Factors of Bacterial Keratitis: The Nottingham Infectious Keratitis Study. *Front. Med.* **2021**, *8*, 715118. [CrossRef] [PubMed]
4. Ting, D.S.; Gopal, B.P.; Deshmukh, R.; Seitzman, G.D.; Said, D.G.; Dua, H.S. Diagnostic armamentarium of infectious keratitis: A comprehensive review. *Ocul. Surf.* **2022**, *23*, 27–39. [CrossRef]
5. Ung, L.; Bispo, P.J.; Shanbhag, S.S.; Gilmore, M.S.; Chodosh, J. The persistent dilemma of microbial keratitis: Global burden, diagnosis, and antimicrobial resistance. *Surv. Ophthalmol.* **2019**, *64*, 255–271. [CrossRef]
6. Thomas, P.A.; Geraldine, P. Infectious keratitis. *Curr. Opin. Infect. Dis.* **2007**, *20*, 129–141. [CrossRef]
7. Hafezi, F.; Tabibian, D.; Richoz, O. PACK-CXL: Corneal cross-linking for treatment of infectious keratitis. *J. Ophthalmic Vis. Res.* **2015**, *10*, 77–80. [CrossRef]
8. Dalmon, C.; Porco, T.C.; Lietman, T.M.; Prajna, N.V.; Prajna, L.; Das, M.R.; Kumar, J.A.; Mascarenhas, J.; Margolis, T.P.; Whitcher, J.P.; et al. The Clinical Differentiation of Bacterial and Fungal Keratitis: A Photographic Survey. *Investig. Ophthalmol. Vis. Sci.* **2012**, *53*, 1787–1791. [CrossRef]
9. Tandon, R.; Gupta, N. Investigative modalities in infectious keratitis. *Indian J. Ophthalmol.* **2008**, *56*, 209. [CrossRef]
10. Mantopoulos, D.; Cruzat, A.; Hamrah, P. In Vivo Imaging of Corneal Inflammation: New Tools for Clinical Practice and Research. *Semin. Ophthalmol.* **2010**, *25*, 178–185. [CrossRef]

11. Lin, A.; Rhee, M.K.; Akpek, E.K.; Amescua, G.; Farid, M.; Garcia-Ferrer, F.J.; Varu, D.M.; Musch, D.; Dunn, S.P.; Mah, F.S.; et al. Bacterial Keratitis Preferred Practice Pattern®. *Ophthalmology* **2019**, *126*, P1–P55. [CrossRef] [PubMed]
12. Bourcier, T.; Sauer, A.; Dory, A.; Denis, J.; Sabou, M. Fungal keratitis. *J. Français D'ophtalmol.* **2017**, *40*, e307–e313. [CrossRef] [PubMed]
13. Bamdad, S.; Malekhosseini, H.; Khosravi, A. Ultraviolet A/Riboflavin Collagen Cross-Linking for Treatment of Moderate Bacterial Corneal Ulcers. *Cornea* **2015**, *34*, 402–406. [CrossRef] [PubMed]
14. Skaat, A.; Zadok, D.; Goldich, Y.; Varssano, D.; Berger, Y.; Ezra-Nimni, O.; Avni, I.; Barequet, I.S. Riboflavin/UVA Photochemical Therapy for Severe Infectious Keratitis. *Eur. J. Ophthalmol.* **2013**, *24*, 21–28. [CrossRef] [PubMed]
15. Chan, E.; Snibson, G.R.; Sullivan, L. Treatment of infectious keratitis with riboflavin and ultraviolet-A irradiation. *J. Cataract Refract. Surg.* **2014**, *40*, 1919–1925. [CrossRef] [PubMed]
16. Ferrari, T.M.; Leozappa, M.; Lorusso, M.; Epifani, E.; Ferrari, L.M. *Escherichia Coli* Keratitis Treated with Ultraviolet A/Riboflavin Corneal Cross-Linking: A Case Report. *Eur. J. Ophthalmol.* **2009**, *19*, 295–297. [CrossRef] [PubMed]
17. Al-Sabai, N.; Koppen, C.; Tassignon, M.J. UVA/riboflavin crosslinking as treatment for corneal melting. *Bull. Soc. Belg. Ophtalmol.* **2010**, *315*, 13–17.
18. Erdem, E.; Harbiyeli, I.I.; Boral, H.; Ilkit, M.; Yagmur, M.; Ersoz, R. Corneal Collagen Cross-Linking for the Management of Mycotic Keratitis. *Mycopathologia* **2018**, *183*, 521–527. [CrossRef]
19. Wei, A.; Wang, K.; Wang, Y.; Gong, L.; Xu, J.; Shao, T. Evaluation of corneal cross-linking as adjuvant therapy for the management of fungal keratitis. *Graefes Arch. Clin. Exp. Ophthalmol.* **2019**, *257*, 1443–1452. [CrossRef]
20. Galperin, G.; Berra, M.; Tau, J.; Boscaro, G.; Zarate, J.; Berra, A. Treatment of Fungal Keratitis from Fusarium Infection by Corneal Cross-Linking. *Cornea* **2012**, *31*, 176–180. [CrossRef]
21. Ramona, B.I.; Catalina, C.; Andrei, M.; Daciana, S.; Calin, T. Collagen crosslinking in the management of microbial keratitis. *Rom. J. Ophthalmol.* **2016**, *60*, 28–30. [PubMed]
22. Shetty, R.; Nagaraja, H.; Jayadev, C.; Shivanna, Y.; Kugar, T. Collagen crosslinking in the management of advanced non-resolving microbial keratitis. *Br. J. Ophthalmol.* **2014**, *98*, 1033–1035. [CrossRef] [PubMed]
23. Said, D.G.; Elalfy, M.S.; Gatzioufas, Z.; El-Zakzouk, E.S.; Hassan, M.A.; Saif, M.Y.; Zaki, A.A.; Dua, H.S.; Hafezi, F. Collagen Cross-Linking with Photoactivated Riboflavin (PACK-CXL) for the Treatment of Advanced Infectious Keratitis with Corneal Melting. *Ophthalmology* **2014**, *121*, 1377–1382. [CrossRef] [PubMed]
24. Price, M.O.; Tenkman, L.R.; Schrier, A.; Fairchild, K.M.; Trokel, S.L.; Price, F.W., Jr. Photoactivated Riboflavin Treatment of Infectious Keratitis Using Collagen Cross-linking Technology. *J. Refract. Surg.* **2012**, *28*, 706–713. [CrossRef] [PubMed]
25. Hafezi, F.; Hosny, M.; Shetty, R.; Knyazer, B.; Chen, S.; Wang, Q.; Hashemi, H.; Torres-Netto, E.A.; Zhang, H.; Bora'I, A.; et al. PACK-CXL vs. antimicrobial therapy for bacterial, fungal, and mixed infectious keratitis: A prospective randomized phase 3 trial. *Eye Vis.* **2022**, *9*, 2. [CrossRef]
26. Arance-Gil, Á.; Gutierrez, Á.R.; Villa-Collar, C.; Bona, A.N.; Lopes-Ferreira, D.; González-Méijome, J.M. Corneal cross-linking for Acanthamoeba keratitis in an orthokeratology patient after swimming in contaminated water. *Contact Lens Anterior Eye* **2014**, *37*, 224–227. [CrossRef]
27. Watson, S.H.; Shekhawat, N.S.; Daoud, Y.J. Treatment of recalcitrant Acanthamoeba Keratitis with Photoactivated Chromophore for Infectious Keratitis Corneal Collagen Cross-Linking (PACK-CXL). *Am. J. Ophthalmol. Case Rep.* **2022**, *25*, 101330. [CrossRef]
28. Pettersson, M.N.; Lagali, N.; Mortensen, J.; Jofré, V.; Fagerholm, P. High fluence PACK-CXL as adjuvant treatment for advanced Acanthamoeba keratitis. *Am. J. Ophthalmol. Case Rep.* **2019**, *15*, 100499. [CrossRef]
29. Zamani, M.; Panahi-Bazaz, M.; Assadi, M. Corneal collagen cross-linking for treatment of non-healing corneal ulcers. *J. Ophthalmic Vis. Res.* **2015**, *10*, 16–20. [CrossRef]
30. Sorkhabi, R.; Sedgipoor, M.; Mahdavifard, A. Collagen cross-linking for resistant corneal ulcer. *Int. Ophthalmol.* **2013**, *33*, 61–66. [CrossRef]
31. Panda, A.; Krishna, S.N.; Kumar, S. Photo-Activated Riboflavin Therapy of Refractory Corneal Ulcers. *Cornea* **2012**, *31*, 1210–1213. [CrossRef] [PubMed]
32. Sağlk, A.; Uçakhan, O.; Kanpolat, A. Ultraviolet A and Riboflavin Therapy as an Adjunct in Corneal Ulcer Refractory to Medical Treatment. *Eye Contact Lens* **2013**, *39*, 413–415. [CrossRef] [PubMed]
33. Basaiawmoit, P.; Selvin, S.S.T.; Korah, S. PACK-CXL in Reducing the Time to Heal in Suppurative Corneal Ulcers: Observations of a Pilot Study from South India. *Cornea* **2018**, *37*, 1376–1380. [CrossRef]
34. Tal, K.; Gal-Or, O.; Pillar, S.; Zahavi, A.; Rock, O.; Bahar, I. Efficacy of Primary Collagen Cross-Linking with Photoactivated Chromophore (PACK-CXL) for the Treatment of Staphylococcus aureus–Induced Corneal Ulcers. *Cornea* **2015**, *34*, 1281–1286. [CrossRef] [PubMed]
35. Awad, R.; Hafezi, F.; Ghaith, A.A.; Baddour, M.M.; Awad, K.; Abdalla, M.; Sheta, E.; Sultan, G.M.; Elmassry, A. Comparison between three different high fluence UVA levels in corneal collagen cross-linking for treatment of experimentally induced fungal keratitis in rabbits. *Eur. J. Ophthalmol.* **2022**, *32*, 112067212210922. [CrossRef] [PubMed]
36. Makdoumi, K.; Mortensen, J.; Sorkhabi, O.; Malmvall, B.-E.; Crafoord, S. UVA-riboflavin photochemical therapy of bacterial keratitis: A pilot study. *Graefes Arch. Clin. Exp. Ophthalmol.* **2011**, *250*, 95–102. [CrossRef]
37. Kasetsuwan, N.; Reinprayoon, U.; Satitpitakul, V. Photoactivated Chromophore for Moderate to Severe Infectious Keratitis as an Adjunct Therapy: A Randomized Controlled Trial. *Am. J. Ophthalmol.* **2016**, *165*, 94–99. [CrossRef]

38. Prajna, N.V.; Radhakrishnan, N.; Lalitha, P.; Rajaraman, R.; Narayana, S.; Austin, A.F.; Liu, Z.; Keenan, J.D.; Porco, T.C.; Lietman, T.M.; et al. Cross-Linking Assisted Infection Reduction (CLAIR): A Randomized Clinical Trial Evaluating the Effect of Adjuvant Cross-Linking on Bacterial Keratitis. *Cornea* **2021**, *40*, 837–841. [CrossRef]
39. Zloto, O.; Barequet, I.S.; Weissman, A.; Nimni, O.E.; Berger, Y.; Avni-Zauberman, N. Does PACK-CXL Change the Prognosis of Resistant Infectious Keratitis? *J. Refract. Surg.* **2018**, *34*, 559–563. [CrossRef]
40. Alamillo-Velazquez, J.; Ruiz-Lozano, R.E.; Hernandez-Camarena, J.C.; Rodriguez-Garcia, A. Contact Lens-Associated Infectious Keratitis: Update on Diagnosis and Therapy. In *Infectious Eye Diseases—Recent Advances in Diagnosis and Treatment*; IntechOpen: London, UK, 2021. [CrossRef]
41. Wollensak, G.; Wilsch, M.; Spoerl, E.; Seiler, T. Collagen Fiber Diameter in the Rabbit Cornea After Collagen Crosslinking by Riboflavin/UVA. *Cornea* **2004**, *23*, 503–507. [CrossRef]
42. Wollensak, G.; Spoerl, E.; Seiler, T. Riboflavin/ultraviolet-a–induced collagen crosslinking for the treatment of keratoconus. *Am. J. Ophthalmol.* **2003**, *135*, 620–627. [CrossRef]
43. Perez-Straziota, C.; Gaster, R.N.; Rabinowitz, Y.S. Corneal Cross-Linking for Pediatric Keratcoconus Review. *Cornea* **2018**, *37*, 802–809. [CrossRef] [PubMed]
44. McQuaid, R.; Cummings, A.B.; Mrochen, M. Newer protocols and future in collagen cross-linking. *Indian J. Ophthalmol.* **2013**, *61*, 425–427. [CrossRef] [PubMed]
45. Kumar, V.; Lockerbie, O.; Keil, S.D.; Ruane, P.H.; Platz, M.S.; Martin, C.; Ravanat, J.-L.; Cadet, J.; Goodrich, R. Riboflavin and UV-Light Based Pathogen Reduction: Extent and Consequence of DNA Damage at the Molecular Level. *Photochem. Photobiol.* **2004**, *80*, 15–21. [CrossRef]
46. Spoerl, E.; Wollensak, G.; Seiler, T. Increased resistance of crosslinked cornea against enzymatic digestion. *Curr. Eye Res.* **2004**, *29*, 35–40. [CrossRef]
47. Goodrich, R. The use of riboflavin for the inactivation of pathogens in blood products. *Vox Sang.* **2000**, *78* (Suppl. 2), 211–215.
48. Ruane, P.H.; Edrich, R.; Gampp, D.; Keil, S.D.; Leonard, R.L.; Goodrich, R.P. Photochemical inactivation of selected viruses and bacteria in platelet concentrates using riboflavin and light. *Transfusion* **2004**, *44*, 877–885. [CrossRef]
49. Schnitzler, E.; Spörl, E.; Seiler, T. Irradiation of cornea with ultraviolet light and riboflavin administration as a new treatment for erosive corneal processes, preliminary results in four patients. *Klin. Monbl. Augenheilkd.* **2000**, *217*, 190–193. [CrossRef]
50. Iseli, H.P.; Thiel, M.A.; Hafezi, F.; Kampmeier, J.; Seiler, T. Ultraviolet A/Riboflavin Corneal Cross-linking for Infectious Keratitis Associated with Corneal Melts. *Cornea* **2008**, *27*, 590–594. [CrossRef]
51. Richoz, O.; Kling, S.; Hoogewoud, F.; Hammer, A.; Tabibian, D.; Francois, P.; Schrenzel, J.; Hafezi, F. Antibacterial Efficacy of Accelerated Photoactivated Chromophore for Keratitis–Corneal Collagen Cross-linking (PACK-CXL). *J. Refract. Surg.* **2014**, *30*, 850–854. [CrossRef]
52. Tabibian, D.; Richoz, O.; Riat, A.; Schrenzel, J.; Hafezi, F. Accelerated Photoactivated Chromophore for Keratitis–Corneal Collagen Cross-linking as a First-line and Sole Treatment in Early Fungal Keratitis. *J. Refract. Surg.* **2014**, *30*, 855–857. [CrossRef] [PubMed]
53. Barac, I.R.; Balta, G.; Zemba, M.; Branduse, L.; Mehedintu, C.; Burcea, M.; Barac, D.A.; Branisteanu, D.C.; Balta, F. Accelerated vs. conventional collagen cross-linking for infectious keratitis. *Exp. Ther. Med.* **2021**, *21*, 285. [CrossRef] [PubMed]
54. Knyazer, B.; Krakauer, Y.; Baumfeld, Y.; Lifshitz, T.; Kling, S.; Hafezi, F. Accelerated Corneal Cross-Linking with Photoactivated Chromophore for Moderate Therapy-Resistant Infectious Keratitis. *Cornea* **2018**, *37*, 528–531. [CrossRef] [PubMed]
55. Kling, S.; Hufschmid, F.S.; Torres-Netto, E.; Randleman, J.B.; Willcox, M.; Zbinden, R.; Hafezi, F. High Fluence Increases the Antibacterial Efficacy of PACK Cross-Linking. *Cornea* **2020**, *39*, 1020–1026. [CrossRef] [PubMed]
56. Bilgihan, K.; Kalkanci, A.; Ozdemir, H.B.; Yazar, R.; Karakurt, F.; Yuksel, E.; Otag, F.; Karabicak, N.; Arikan-Akdagli, S. Evaluation of Antifungal Efficacy of 0.1% and 0.25% Riboflavin with UVA: A Comparative In Vitro Study. *Curr. Eye Res.* **2016**, *41*, 1050–1056. [CrossRef]
57. Awad, E.A.; Abdelkader, M.; Abdelhameed, A.G.; Gaafar, W.; Mokbel, T.H. Collagen crosslinking with photoactivated riboflavin in advanced infectious keratitis with corneal melting: Electrophysiological Study. *Int. J. Ophthalmol.* **2020**, *13*, 574–579. [CrossRef] [PubMed]
58. Deshmukh, R. Commentary: PACK-CXL in fungal keratitis. *Indian J. Ophthalmol.* **2019**, *67*, 1701–1702. [CrossRef]
59. Makdoumi, K.; Bäckman, A. Photodynamic UVA-riboflavin bacterial elimination in antibiotic-resistant bacteria. *Clin. Exp. Ophthalmol.* **2016**, *44*, 582–586. [CrossRef]
60. Ting, D.S.J.; Henein, C.; Said, D.G.; Dua, H.S. Photoactivated chromophore for infectious keratitis—Corneal cross-linking (PACK-CXL): A systematic review and meta-analysis. *Ocul. Surf.* **2019**, *17*, 624–634. [CrossRef]
61. Davis, S.A.; Bovelle, R.; Han, G.; Kwagyan, J. Corneal collagen cross-linking for bacterial infectious keratitis. *Cochrane Database Syst. Rev.* **2020**, *6*, CD013001. [CrossRef]
62. Chan, T.; Lau, T.W.S.; Lee, J.W.Y.; Wong, I.Y.H.; Jhanji, V.; Wong, R.L.M. Corneal collagen cross-linking for infectious keratitis: An update of clinical studies. *Acta Ophthalmol.* **2015**, *93*, 689–696. [CrossRef]
63. Alio, J.L.; Abbouda, A.; Valle, D.D.; Del Castillo, J.M.B.; Fernandez, J.A.G. Corneal cross linking and infectious keratitis: A systematic review with a meta-analysis of reported cases. *J. Ophthalmic Inflamm. Infect.* **2013**, *3*, 47. [CrossRef] [PubMed]
64. Abbouda, A.; Abicca, I.; Alió, J.L. Current and Future Applications of Photoactivated Chromophore for Keratitis-Corneal Collagen Cross-Linking (PACK-CXL): An Overview of the Different Treatments Proposed. *Semin. Ophthalmol.* **2018**, *33*, 293–299. [CrossRef] [PubMed]

65. Tayapad, J.B.; Viguilla, A.Q.; Reyes, J.M. Collagen cross-linking and corneal infections. *Curr. Opin. Ophthalmol.* **2013**, *24*, 288–290. [CrossRef] [PubMed]
66. Price, M.; Price, F.W. Corneal cross-linking in the treatment of corneal ulcers. *Curr. Opin. Ophthalmol.* **2016**, *27*, 250–255. [CrossRef]
67. Idrus, E.A.; Utti, E.M.; Mattila, J.S.; Krootila, K. Photoactivated chromophore corneal cross-linking (PACK-CXL) for treatment of severe keratitis. *Acta Ophthalmol.* **2019**, *97*, 721–726. [CrossRef]
68. Rosetta, P.; Legrottaglie, E.F.; Pagano, L.; Vinciguerra, P. Corneal Cross-Linking Window Absorption (CXL-WA) as an Adjuvant Therapy in the Management of *Aspergillus niger* Keratitis. *Case Rep. Ophthalmol. Med.* **2018**, *2018*, 4856019. [CrossRef]
69. Vajpayee, R.B.; Shafi, S.N.; Maharana, P.K.; Sharma, N.; Jhanji, V. Evaluation of corneal collagen cross-linking as an additional therapy in mycotic keratitis. *Clin. Exp. Ophthalmol.* **2015**, *43*, 103–107. [CrossRef]
70. Yagci, A.; Palamar, M.; Hilmioglu, S.P.; Irkec, M. Cross-Linking Treatment and Corneal Transplant in Refractory Acremonium Keratitis: Case Report. *Exp. Clin. Transplant.* **2016**, *14*, 580–583.
71. Uddaraju, M.; Mascarenhas, J.; Das, M.R.; Radhakrishnan, N.; Keenan, J.D.; Prajna, L.; Prajna, V.N. Corneal Cross-linking as an Adjuvant Therapy in the Management of Recalcitrant Deep Stromal Fungal Keratitis: A Randomized Trial. *Am. J. Ophthalmol.* **2015**, *160*, 131–134.e5. [CrossRef]
72. Prajna, N.V.; Radhakrishnan, N.; Lalitha, P.; Austin, A.; Ray, K.J.; Keenan, J.D.; Porco, T.C.; Lietman, T.M.; Rose-Nussbaumer, J. Cross-Linking–Assisted Infection Reduction. *Ophthalmology* **2019**, *127*, 159–166. [CrossRef] [PubMed]
73. Singhal, D.; Maharana, P.K. Re: Prajna et al.: Cross-Linking-Assisted Infection Reduction: A randomized clinical trial evaluating the effect of adjuvant cross-linking on outcomes in fungal keratitis (Ophthalmology. 2020;127:159–166). *Ophthalmology* **2021**, *128*, e4–e5. [CrossRef] [PubMed]
74. Ting, D.S.J.; Henein, C.; Said, D.G.; Dua, H.S. Re: Prajna et al.: Cross-Linking-Assisted Infection Reduction (CLAIR): A randomized clinical trial evaluating the effect of adjuvant cross-linking on outcomes in fungal keratitis (Ophthalmology. 2020;127:159-166). *Ophthalmology* **2020**, *127*, e55–e56. [CrossRef]
75. Garg, P.; Das, S.; Roy, A. Collagen Cross-linking for Microbial Keratitis. *Middle East Afr. J. Ophthalmol.* **2017**, *24*, 18–23. [CrossRef] [PubMed]
76. Safety and Efficacy Study of Corneal Collagen Cross-Linking in Eyes with Keratoconus. Available online: https://clinicaltrials.gov/ct2/show/study/NCT01344187 (accessed on 27 March 2022).

Journal of *Personalized Medicine*

MDPI

Article

An Outperforming Artificial Intelligence Model to Identify Referable Blepharoptosis for General Practitioners

Ju-Yi Hung [1,2,†], Ke-Wei Chen [1,3,†], Chandrashan Perera [1], Hsu-Kuang Chiu [4], Cherng-Ru Hsu [5], David Myung [1], An-Chun Luo [6], Chiou-Shann Fuh [2], Shu-Lang Liao [7,8,*] and Andrea Lora Kossler [1,*]

1 Ophthalmology, Byers Eye Institute, Stanford University School of Medicine, 2452 Watson Court, Palo Alto, CA 94303, USA; hungjuyi@gmail.com (J.-Y.H.); gosienna@gmail.com (K.-W.C.); chandrashan@gmail.com (C.P.); david.myung@stanford.edu (D.M.)
2 Department of Computer Science and Information Engineering, National Taiwan University, Taipei 10617, Taiwan; fuh@csie.ntu.edu.tw
3 Department of Biomedical Engineering, National Cheng Kung University, Tainan City 70101, Taiwan
4 Computer Science, Stanford University, Stanford, CA 94305, USA; hsu-kuang.chiu@alumni.stanford.edu
5 Ophthalmology, Tri-Service General Hospital, National Defense Medical Center, Taipei 114, Taiwan; josephinesheu@gmail.com
6 Department of Electronic and Optoelectronic System Research Laboratories, Industrial Technology Research Institute, Hsinchu 31040, Taiwan; anchunluo@itri.org.tw
7 Ophthalmology, National Taiwan University Hospital, Taipei 100, Taiwan
8 College of Medicine, National Taiwan University, Taipei 10617, Taiwan
* Correspondence: liaosl89@ntu.edu.tw (S.-L.L.); akossler@stanford.edu (A.L.K.)
† These authors contributed equally to this work.

Abstract: The aim of this study is to develop an AI model that accurately identifies referable blepharoptosis automatically and to compare the AI model's performance to a group of non-ophthalmic physicians. In total, 1000 retrospective single-eye images from tertiary oculoplastic clinics were labeled by three oculoplastic surgeons as having either ptosis, including true and pseudoptosis, or a healthy eyelid. A convolutional neural network (CNN) was trained for binary classification. The same dataset was used in testing three non-ophthalmic physicians. The CNN model achieved a sensitivity of 92% and a specificity of 88%, compared with the non-ophthalmic physician group, which achieved a mean sensitivity of 72% and a mean specificity of 82.67%. The AI model showed better performance than the non-ophthalmic physician group in identifying referable blepharoptosis, including true and pseudoptosis, correctly. Therefore, artificial intelligence-aided tools have the potential to assist in the diagnosis and referral of blepharoptosis for general practitioners.

Keywords: artificial intelligence; blepharoptosis; general practitioners; computer-aided diagnosis (CAD)

Citation: Hung, J.-Y.; Chen, K.-W.; Perera, C.; Chiu, H.-K.; Hsu, C.-R.; Myung, D.; Luo, A.-C.; Fuh, C.-S.; Liao, S.-L.; Kossler, A.L. An Outperforming Artificial Intelligence Model to Identify Referable Blepharoptosis for General Practitioners. *J. Pers. Med.* **2022**, *12*, 283. https://doi.org/10.3390/jpm12020283

Academic Editor: Chieh-Chih Tsai

Received: 28 December 2021
Accepted: 6 February 2022
Published: 15 February 2022

1. Introduction

Blepharoptosis, also known as ptosis, is the drooping or inferior displacement of the upper eyelid. Ptosis can obstruct the visual axis and affect vision and can be a presenting sign of a serious medical disorder, such as ocular myasthenia [1], third cranial nerve palsy [2], or Horner syndrome [3]. It is important for general practitioners to accurately diagnosis ptosis to assist in decision making for referral and work up when necessary. Ptosis is diagnosed by using a ruler and light source to measure the distance between the pupillary light reflex and the upper eyelid margin (margin reflex distance 1, or MRD1) with the eyes in the primary position [4]. With low repeatability and reproducibility in measuring eyelid landmarks and the effect of learning curves [5,6], accurately recognizing ptosis is challenging especially for non-ophthalmologists. Therefore, an automated tool for ptosis diagnosis may be useful for general practitioners.

Currently, artificial intelligence (AI)-aided diagnostic tools play a promising role in the automatic detection of certain diseases, such as diabetic retinopathy [7] and skin cancer [8] from retinal fundus and skin images, respectively. Convolutional neural network (CNN)-based deep learning methods, a subset of machine learning techniques, have been the state of the art in AI for years, leading to enhanced performance in various medical applications [9]. It requires less supervision and uses an end-to-end learning mechanism to map raw inputs, such as image pixels, to outputs without human-directed manipulation of data [10]. The image-to-classification approach in one classifier replaces the multiple steps of previous image analysis methods [11].

In a previous study [12], a variety of CNN architectures, such as VGG-16 [13], ResNet [14], and DenseNet [15], diagnosed true blepharoptosis without any inputs of eyelid measurements from a clinical photograph, achieving a high accuracy of 83.3% to 88.6%. In this study, we further trained an AI model using the VGG-16 architecture with larger and more diverse datasets to accurately diagnose blepharoptosis and compared the AI model's performance to a group of non-ophthalmic physicians. Our goal was to determine if our AI model could outperform physicians to support the need for an AI tool to diagnose blepharoptosis.

2. Materials and Methods

2.1. Image Preparation

Original photographs, taken by a hand-held digital camera (Canon DIGITAL IXUS 950 IS) at a tertiary oculoplastic clinic of adult patients over 20 years old, were retrospectively collected over the past 20 years for surgical evaluation. A total of 1000 images were used in this study. IRB approval was granted for this study by Stanford University, and the research was conducted in accordance with National Taiwan University IRB protocol.

In order to crop a standardized image of a single eye, OpenFace [16], an open-source package, was utilized to identify major facial landmarks in each photograph. Cropped single-eye images were 400 × 600 pixels individually and were then resized to 200 × 300 pixels, matching the input size, which was ready to be used in the CNN architectures.

2.2. Inclusion and Exclusion

After cropping, the photographs involved only the periocular region of a single eyes Figure 1. The appearance of a healthy eyelid is illustrated in Figure 1a. The referable ptosis group included mild ptosis, severe ptosis, and pseudoptosis (dermatochalasis), a condition in which excess upper eyelid skin overhangs the eyelid margin Figure 1b. Upper eyelid retraction was excluded Figure 1c. Poor quality images, including uncentered visual fixation, uneven curves of the upper eyelids, and blurred upper eyelid margins due to dense eyelashes, were excluded. A total of 1000 images were evaluated and 218 images were removed, leaving 782 images for use in this study.

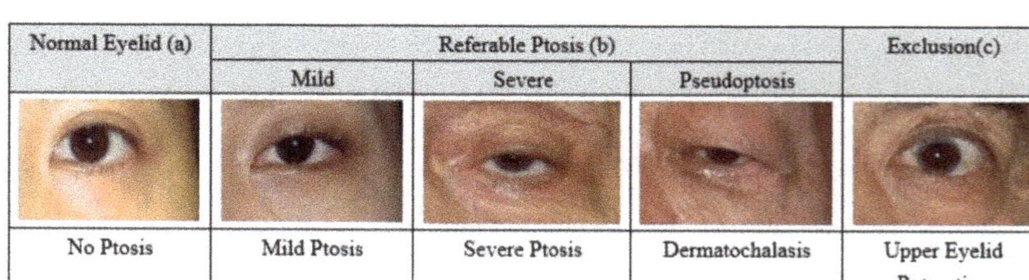

Figure 1. Healthy (**a**), referable ptosis (**b**), and excluded (**c**) group (right eyes).

The brow region was not included in the photographs; therefore, brow ptosis was not excluded. Exact measurements, such as margin to reflex distance 1 (MRD-1), MRD-2 [17,18], levator function [19], or palpebral aperture [20], were not provided. The condition of the other eye and the history of the patients were withheld.

2.3. Annotations for the Ground Truth

Two labelers, both oculoplastic surgeons, achieved an 82% consensus rate in discussion meetings. The major reasons for their disagreements were decisions about healthy eyelids and mild ptosis. To lessen spectrum bias, a third senior oculoplastic surgeon, as an arbiter, yielded the decisive answer for these disagreements, which included 24 images. Figure 2 shows the voting system, with 593 images (accounting for 75%) in the referable ptosis group and 189 (25%) images in the healthy group.

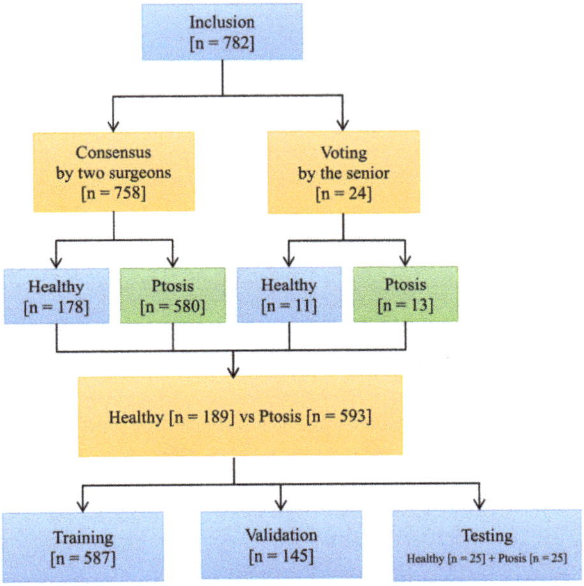

Figure 2. Flowchart for data labeling.

2.4. Data Allocation for Training, Validation, and Testing

A total of 50 images, including 25 healthy eyelids and 25 ptotic eyelids, were randomly selected into testing datasets. The same testing datasets were used to test the AI model and the physician group. The rest of the photographs were then divided into training and validation datasets with the ratio of 8:2 Table 1.

Table 1. The number of images in the training, validation, and testing set.

	Training	Validation (for Training)	Testing
Referable ptosis group	455	113	25
Healthy group	132	32	25

2.5. Model Architecture and Training

VGG-16 was used as the base structure [13,21]. The last few layers of VGG-16's architecture were replaced with a global max pooling layer followed by fully connected layers and a sigmoid function for our binary classification problem. In order to reduce memory usage, the size of the input images was adjusted to 200 × 300 pixels. The details of our model architecture can be seen in Table 2.

Table 2. Structure of the model.

Input Size	Layer	Output Size	Number of Feature Maps	Kernel Size	Stride	Activation
-	Image	200 × 300 × 3	-	-	-	-
200 × 300 × 3	Convolution	200 × 300 × 64	64	3 × 3	1	ReLU
200 × 300 × 64	Convolution	200 × 300 × 64	64	3 × 3	1	ReLU
200 × 300 × 64	Max pooling	100 × 150 × 64	64	-	2	-
100 × 150 × 64	Convolution	100 × 150 × 128	128	3 × 3	1	ReLU
100 × 150 × 128	Convolution	100 × 150 × 128	128	3 × 3	1	ReLU
100 × 150 × 128	Max pooling	50 × 75 × 128	128	-	2	-
50 × 75 × 128	Convolution	50 × 75 × 256	256	3 × 3	1	ReLU
50 × 75 × 256	Convolution	50 × 75 × 256	256	3 × 3	1	ReLU
50 × 75 × 256	Global max pooling	1 × 256	-	-	-	-
1 × 256	Fully connected	1 × 512	-	-	-	ReLU
1 × 512	Fully connected	1	-	-	-	Sigmoid

2.6. Transfer Learning and Data Augmentation

Transfer learning was performed by importing weights trained on ImageNet [22]. Tensorflow 2.0 with Keras was used as our training framework. For learning rate optimization, Adam optimizer was applied [21]. Data augmentation was also used to prevent overfitting. The transformations of photographs included:

- Images flipped horizontally;
- Random image rotations of up to 15 degrees;
- Random zooms in or out between the range of 90% to 120%;
- Adjusted brightness/contrast by 50%;
- Images shifted horizontally or vertically by 10%.

2.7. Testing in Non-Ophthalmic Physician Group

Three specialists, one each from emergency medicine, neurology, and family medicine, were tested on behalf of the non-ophthalmic physician group. The clinical experience of each of the three physicians was over five years. The same testing set, including 25 healthy eyelids and 25 ptotic eyelids, was given to the group to distinguish ptotic eyelids from healthy eyelids. No other information, such as MRD-1 measurements, the condition of the other eye, or patient histories, were provided. Moreover, no further training on blepharoptosis diagnosis was given. The decision making relied on each physician's personal background knowledge.

3. Results

There were 45 correct predictions, including 22 healthy and 23 ptosis answers, by the CNN model from a total of 50 testing images. The accuracy of the AI model was 90%, with a sensitivity of 92% and a specificity of 88%. Three false positives and two false negatives were found.

3.1. Confusion Matrix and ROC Curve

The confusion matrix with a 0.5 threshold setting is shown in Figure 3. The receiver operating characteristic (ROC) curve is presented in Figure 4. The area under the curve (AUC) was 0.987. The mean accuracy of the non-ophthalmic physician group was 77.33% (range: 70–82%) with a mean sensitivity of 72% (range: 68–76%) and a mean specificity of 82.67% (range: 72–88%), as seen in Figure 5.

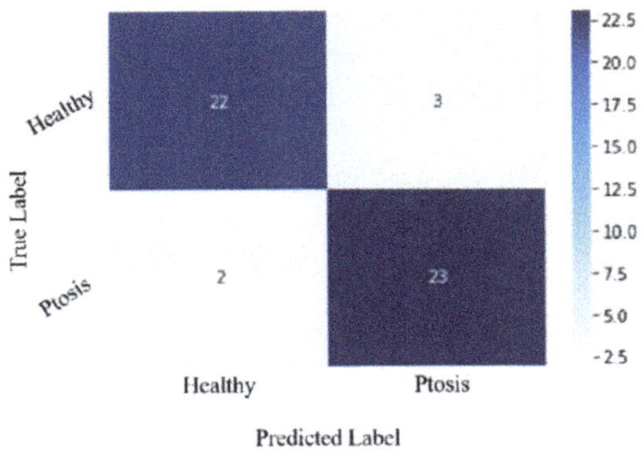

Figure 3. Confusion matrix. The threshold is 0.5.

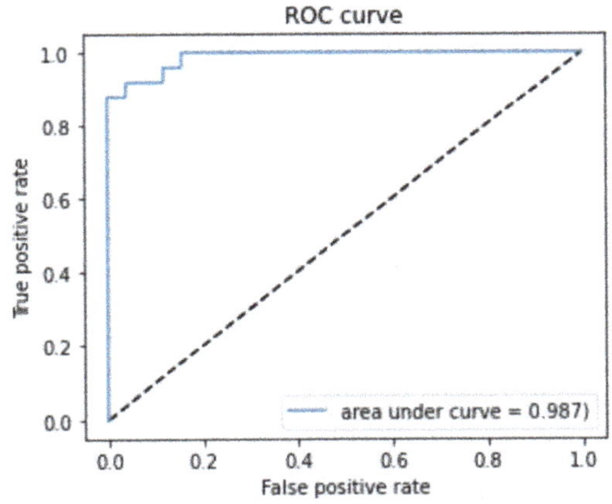

Figure 4. ROC curve. The area under the curve (AUC) is 0.987.

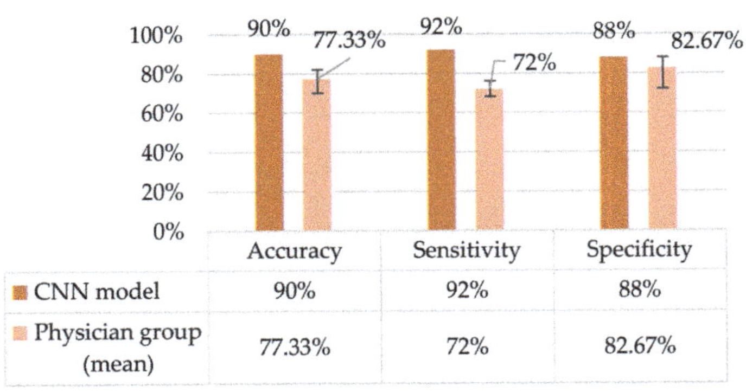

Figure 5. Performance comparison.

3.2. Grad-CAM Results

Gradient-weighted class activation mapping (Grad-CAM) [23] was applied to visualize the AI model. The result showed that the weight in the background was around 0~0.2. In the ptotic eyelids, the area between the upper eyelid margin and the central cornea light reflex showed the highest weight, around 0.5~1.0 (Figure 6).

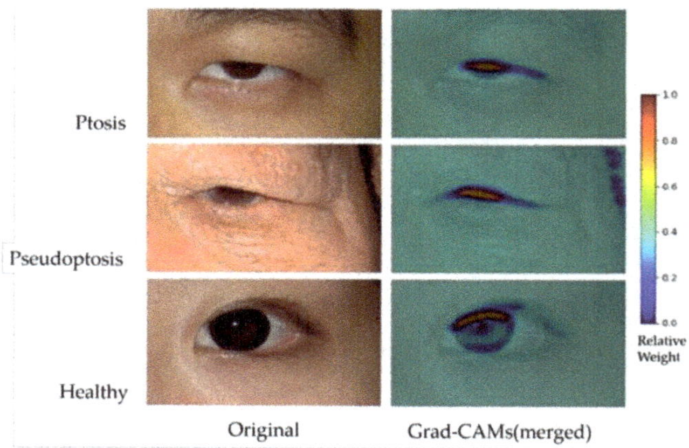

Figure 6. Original images and Grad-CAM results of the AI model predictions. Ptosis (**upper**), pseudoptosis (**middle**), and healthy eyelids (**lower**). Grad-CAM results have been merged with original images.

4. Discussion

It is important for general practitioners to promptly diagnose and refer eyelid ptosis, including pseudoptosis, to ophthalmic specialists for further evaluation, work up, and treatment. Pseudoptosis is a heterogeneous group of disorders where the upper eyelid can drop in the absence of pathology of the upper eyelid muscles [24]. Dermatochalasis is likely the most common eyelid condition that causes confusion when evaluating a patient with apparent ptosis. Excess upper eyelid skin may overhang the eyelashes and obstruct the visualization of the eyelid margin, giving the impression of a low-lying eyelid. In a previous proof-of-concept study, we demonstrated that an AI model could detect true ptosis from healthy eyelids [12]. In this study, we evaluated true ptosis and pseudoptosis versus health eyelids, applied a larger dataset of 782 images, and compared the AI model performance to non-ophthalmic physicians. Our results demonstrate that the AI model achieved an accuracy of 90%, with 92% specificity and 88% sensitivity. Additionally, the AI model performed well even when including pseudoptosis cases, which better mimic the real clinical situation in primary care.

A non-ophthalmic group of three physicians, including experts in family medicine, neurology, and emergency medicine, were chosen as a comparator group. The family medicine doctor represented general practitioners who are commonly the first line in seeing and diagnosing age-related and systemic causes of ptosis. The neurologist was selected due to specialized training in diagnosing ptosis, particularly related to neurologic or myogenic causes. Finally, the emergency medicine doctor was selected due to expertise in diagnosing acute causes of ptosis, such as Horner syndrome, third nerve palsy [25], or trauma. Hence, our non-ophthalmic group had previous experience in identifying blepharoptosis. Our results demonstrate a mean accuracy of 77.33% (range: 70–82%), with a mean sensitivity of 72% (range: 68–76%) and a mean specificity of 82.67% (range: 72–88%) in the non-ophthalmic physician group, while the AI model achieved an accuracy of 90%, with a sensitivity of 92% and a specificity of 88%. These results suggest that an AI-aided diagnostic tool can accurately detect blepharoptosis and prompt referral for ophthalmic evaluation when necessary.

CNNs (convolutional neural networks) have achieved great success in image classification. For example, in the current largest image classification dataset classification challenge, ImageNet, all models with top performance used CNN architectures. The general trend is that the deeper the model, the greater discernment the model can provide. Some model structures can be very deep, such as Res Net-152, which has 152 CNN layers. In a previous smaller scale study [12], where less than 500 eyelid pictures were evaluated, a variety of CNN models demonstrated high performance over an accuracy of 80%. This study showed that, for ptosis classification, the most common models, such as VGG, ResNet, AlexNet, SqueezeNet, and DenseNet, all have similar performance. Therefore, among those models, we chose a relatively simple model, VGG-16, a computing resource-efficient model, as our base model. The VGG-16 model is based on architecture developed by the Oxford Visual Geometry Group (VGG) and achieved top performance in the ImageNet Large Scale Visual Recognition Challenge (ILSVRC) 2014.

Gradient-weighted class activation mapping (Grad-CAM) is a visual explanation of the AI model, which is applicable to a wide variety of CNN model-families [23]. To aid the understanding of AI model predictions, a heat map identifies the areas of the input image that contributed most to the AI model's classification using a technique called class activation mappings. In addition, to visualize reasonable AI predictions, Grad-CAM explanations also helped identify dataset biases in images. For example, a preoperative marking around the eye or a postoperative suture on the eyelid may provide misleading clues to the AI model, rather than eyelid information for blepharoptosis. The results of Grad-CAM (Figure 6) demonstrated a hotspot area (0.5–1.0 in weights) between the upper eyelid margin and central corneal light reflex, which is clinically compatible with the MRD-1 concept. The cold zone (0–0.2 in weights) in the background successfully excluded dataset biases, providing stronger faithfulness. With larger and more diverse data utilization in the future, more precise results to understand the AI predictions can be expected.

AI-assisted ptosis diagnostic tools can be of great impact on the management of congenital ptosis, since up to one-third of congenital ptosis patients are at risk for amblyopia [26]. The accurate diagnosis of ptosis based on external photographs would prove especially helpful in the pediatric population for ophthalmologists and general practitioners alike, as the eyelid exam can be challenging in uncooperative or crying children, patients with developmental delays, and babies. The AI-assisted detection of congenital ptosis could have a huge impact on preventing and treating amblyopia promptly. External validation with outsourced images, including mobile phone photographs, to confirm the strength and weakness of this AI model also deserves further investigation.

Limitations to this study include that the data resource was only from Asian ethnicities, setting limitations in both model training and testing process. Future studies will analyze external photographs from diverse ethnicities to further train the AI model and expand the application for all users. Additionally, only adults were included in this study, setting limitations for pediatric care. Furthermore, we did not measure variables including palpebral aperture, levator muscle excursion, and brow position, which should be identified for detailed and quantifiable ptosis assessment.

There were also inherent limitations in the labeling of the ground truths by three oculoplastic surgeons. Some photos of mild ptosis were challenging to differentiate from normal eyelids, even among experienced oculoplastic surgeons. Hence, the AI model constructed in our study only provided information as to whether a photo might have ptosis from an oculoplastic surgeon's point of view. This might also explain why this study did not achieve much greater accuracy then our previous study, since more data may introduce more photos with uncertainty [12].

5. Conclusions

The AI model using CNNs achieved better performance than the non-ophthalmic physician group and shows value as a diagnostic tool to be used in assisting the referral of blepharoptosis, including true and pseudoptosis.

Author Contributions: Conceptualization, J.-Y.H. and K.-W.C.; methodology, J.-Y.H., K.-W.C., and C.P.; software, K.-W.C.; validation, J.-Y.H. and C.-R.H. ; formal analysis, J.-Y.H.; investigation, J.-Y.H.; resources, D.M., A.-C.L., and C.-S.F.; data curation, S.-L.L.; writing—original draft preparation, J.-Y.H. and K.-W.C.; writing—review and editing, C.P. and H.-K.C.; visualization, K.-W.C.; supervision, S.-L.L. and A.L.K.; project administration, S.-L.L. and A.L.K. All authors have read and agreed to the published version of the manuscript.

Funding: Authors ALK and DM were supported by departmental core grants from Research to Prevent Blindness and the National Eye Institute (P30-026877).

Institutional Review Board Statement: The study was conducted in accordance with the Declaration of Helsinki, and approved by the 121st meeting of the Research Ethics Committee D of the National Taiwan University Hospital (protocol code: NTUH-REC No. 201908066RIND, and date of approval: 12 November 2021).

Informed Consent Statement: Our IRB-approved data and the request for the waiver of patient informed consent have been approved by the 121st meeting of Research Ethics Committee D of the National Taiwan University Hospital on 12 November 2021. The reason for waiver of patient informed consent is that the data is the retrospective large scale of single-eye-only regional images, involving over 1000 cases, which were applied de-identification methods to remove all 18 identifiers, in accordance with the Health Insurance Portability and Accountability Act (HIPAA) Privacy Rule, or Standards for Privacy of Individually Identifiable Health Information, complying with standards for the protection of certain health information by the U.S. Department of Health and Human Services (HHS).

Data Availability Statement: The data presented in this study are available on request from the corresponding author. The data are not publicly available due to privacy issue.

Acknowledgments: We thank Taiwan National Center for High-performance Computing (NCHC) for providing computational and storage resources, departmental core grants from the National Eye Institute (P30 EY026877) and Research to Prevent Blindness (RPB) to the Byers Eye Institute at Stanford, and Karen Chang for assisting in the research.

Conflicts of Interest: The authors declare no conflict of interest.

References

1. Al-Haidar, M.; Benatar, M.; Kaminski, H.J. Ocular Myasthenia. *Neurol. Clin.* **2018**, *36*, 241–251. [CrossRef] [PubMed]
2. Bagheri, A.; Borhani, M.; Salehirad, S.; Yazdani, S.; Tavakoli, M. Blepharoptosis Associated With Third Cranial Nerve Palsy. *Ophthalmic Plast. Reconstr. Surg.* **2015**, *31*, 357–360. [CrossRef] [PubMed]
3. Martin, T.J. Horner Syndrome: A Clinical Review. *ACS Chem. Neurosci.* **2018**, *9*, 177–186. [CrossRef] [PubMed]
4. Putterman, A.M. Margin reflex distance (MRD) 1, 2, and 3. *Ophthalmic Plast. Reconstr. Surg.* **2012**, *28*, 308–311. [CrossRef]
5. Boboridis, K.; Assi, A.; Indar, A.; Bunce, C.; Tyers, A. Repeatability and reproducibility of upper eyelid measurements. *Br. J. Ophthalmol.* **2001**, *85*, 99–101. [CrossRef]
6. Nemet, A.Y. Accuracy of marginal reflex distance measurements in eyelid surgery. *J. Craniofacial Surg.* **2015**, *26*, e569–e571. [CrossRef]
7. Gulshan, V.; Peng, L.; Coram, M.; Stumpe, M.C.; Wu, D.; Narayanaswamy, A.; Venugopalan, S.; Widner, K.; Madams, T.; Cuadros, J.; et al. Development and Validation of a Deep Learning Algorithm for Detection of Diabetic Retinopathy in Retinal Fundus Photographs. *JAMA* **2016**, *316*, 2402–2410. [CrossRef]
8. Esteva, A.; Kuprel, B.; Novoa, R.A.; Ko, J.; Swetter, S.M.; Blau, H.M.; Thrun, S. Dermatologist-level classification of skin cancer with deep neural networks. *Nature* **2017**, *542*, 115–118. [CrossRef]
9. Shen, D.; Wu, G.; Suk, H.-I. Deep Learning in Medical Image Analysis. *Annu. Rev. Biomed. Eng.* **2017**, *19*, 221–248. [CrossRef]
10. Wang, F.; Casalino, L.P.; Khullar, D. Deep learning in medicine—promise, progress, and challenges. *JAMA internal medicine* **2019**, *179*, 293–294. [CrossRef]
11. Zeiler, M.D.; Fergus, R. *Visualizing and Understanding Convolutional Networks*; Springer: Cham, Switzerland, 2014; pp. 818–833.
12. Hung, J.-Y.; Perera, C.; Chen, K.-W.; Myung, D.; Chiu, H.-K.; Fuh, C.-S.; Hsu, C.-R.; Liao, S.-L.; Kossler, A.L. A deep learning approach to identify blepharoptosis by convolutional neural networks. *Int. J. Med. Inform.* **2021**, *148*, 104402. [CrossRef] [PubMed]
13. Simonyan, K.; Zisserman, A. Very deep convolutional networks for large-scale image recognition. *arXiv preprint* **2014**, arXiv:1409.1556.
14. He, K.; Zhang, X.; Ren, S.; Sun, J. Deep residual learning for image recognition. In *Proceedings of the Proceedings of the IEEE conference on computer vision and pattern recognition, Las Vegas, NV, USA, 27–30 June 2016*; pp. 770–778.

15. Huang, G.; Liu, Z.; Van Der Maaten, L.; Weinberger, K.Q. Densely connected convolutional networks. In Proceedings of the IEEE Conference on Computer Vision and Pattern Recognition (CVPR), Honolulu, HI, USA, 21–26 July 2017; pp. 2261–2269.
16. Amos, B.; Ludwiczuk, B.; Satyanarayanan, M. Openface: A general-purpose face recognition library with mobile applications. *CMU Sch. Comput. Sci.* **2016**, *6*.
17. Bodnar, Z.M.; Neimkin, M.; Holds, J.B. Automated Ptosis Measurements From Facial Photographs. *JAMA Ophthalmol.* **2016**, *134*, 146–150. [CrossRef]
18. Lou, L.; Yang, L.; Ye, X.; Zhu, Y.; Wang, S.; Sun, L.; Qian, D.; Ye, J. A Novel Approach for Automated Eyelid Measurements in Blepharoptosis Using Digital Image Analysis. *Curr. Eye Res.* **2019**, *44*, 1075–1079. [CrossRef]
19. Lai, H.-T.; Weng, S.-F.; Chang, C.-H.; Huang, S.-H.; Lee, S.-S.; Chang, K.-P.; Lai, C.-S. Analysis of Levator Function and Ptosis Severity in Involutional Blepharoptosis. *Ann. Plast. Surg.* **2017**, *78*, S58–S60. [CrossRef]
20. Thomas, P.B.; Gunasekera, C.D.; Kang, S.; Baltrusaitis, T. An Artificial Intelligence Approach to the Assessment of Abnormal Lid Position. *Plast. Reconstr. Surg. Glob. Open* **2020**, *8*, e3089. [CrossRef]
21. Da, K. A method for stochastic optimization. *arXiv preprint* **2014**, arXiv:1412.6980.
22. Russakovsky, O.; Deng, J.; Su, H.; Krause, J.; Satheesh, S.; Ma, S.; Huang, Z.; Karpathy, A.; Khosla, A.; Bernstein, M. Imagenet large scale visual recognition challenge. *Int. J. Comput. Vis.* **2015**, *115*, 211–252. [CrossRef]
23. Selvaraju, R.R.; Cogswell, M.; Das, A.; Vedantam, R.; Parikh, D.; Batra, D. Grad-cam: Visual explanations from deep networks via gradient-based localization. In Proceedings of the IEEE International Conference on Computer Vision (ICCV), Venice, Italy, 22–29 October 2017; pp. 618–626.
24. Cohen, A.J.; Weinberg, D.A. Pseudoptosis. In *Evaluation and Management of Blepharoptosis*; Springer: New York, NY, USA, 2011.
25. Huff, J.S.; Austin, E.W. Neuro-Ophthalmology in Emergency Medicine. *Emerg. Med. Clin. N. Am.* **2016**, *34*, 967–986. [CrossRef]
26. Zhang, J.Y.; Zhu, X.W.; Ding, X.; Lin, M.; Li, J. Prevalence of amblyopia in congenital blepharoptosis: A systematic review and Meta-analysis. *Int. J. Ophthalmol.* **2019**, *12*, 1187–1193. [CrossRef] [PubMed]

Journal of *Personalized Medicine*

MDPI

Article

Long-Term Follow-Up in IgG4-Related Ophthalmic Disease: Serum IgG4 Levels and Their Clinical Relevance

Wei-Yi Chou [1], Ching-Yao Tsai [1,2,3,4] and Chieh-Chih Tsai [5,6,*]

1 Department of Ophthalmology, Zhongxing Branch, Taipei City Hospital, Taipei 103212, Taiwan
2 Institute of Public Health, National Yang Ming Chiao Tung University, Taipei 11221, Taiwan
3 Department of Business Administration, Fu Jen Catholic University, New Taipei City 24205, Taiwan
4 General Education Center, University of Taipei, Taipei 111036, Taiwan
5 Department of Ophthalmology, Taipei Veterans General Hospital, Taipei 11217, Taiwan
6 School of Medicine, National Yang Ming Chiao Tung University, Taipei 11221, Taiwan
* Correspondence: cctsai@vghtpe.gov.tw; Tel.: +886-2-28757325

Abstract: (1) Background: To analyze the association between long-term changes in serum IgG4 levels and the clinical course of patients with IgG4-related ophthalmic disease (IgG4-ROD). (2) Methods: Retrospective analysis of 25 patients with IgG4-ROD. (3) Results: Mean age at diagnosis was 60.68 years. Fifty-six percent of patients had bilateral ocular involvement and 32% had systemic associations. The ocular structures involved were the lacrimal gland (76%), orbital soft tissue (36%), extraocular muscle (20%) and infraorbital nerve (20%). According to last follow-up, 9 (36%) patients had normalized IgG4 levels, and 16 (64%) patients had elevated IgG4 levels. Patients with normalized IgG4 levels had better response to initial steroid treatment and attained a significantly lower IgG4 level after treatment ($p = 0.002$). The highest IgG4 levels were at baseline and disease recurrence, and lowest after initial treatment. At final follow-up, IgG4 levels differed in patients with remission (mean 326.25 mg/dL) and stable disease (mean 699.55 mg/dL). Subgroup analysis was performed in patients with remission, categorized according to whether IgG4 levels were normalized (9 patients) or elevated (10 patients) on last follow up. The elevated group had a higher percentage of bilateral disease, lacrimal gland involvement and recurrence. (4) Conclusions: IgG4-ROD patients with a greater response to initial steroid therapy were more inclined to have normalized IgG4 levels in the long term. Some patients remained in remission despite persistently elevated IgG4 levels, and had regular follow-up without treatment.

Keywords: IgG4-related ophthalmic disease; IgG4-related disease; serum IgG4; corticosteroids

Citation: Chou, W.-Y.; Tsai, C.-Y.; Tsai, C.-C. Long-Term Follow-Up in IgG4-Related Ophthalmic Disease: Serum IgG4 Levels and Their Clinical Relevance. *J. Pers. Med.* **2022**, *12*, 1963. https://doi.org/12121963

Academic Editor: Mirosław J. Szczepański

Received: 16 September 2022
Accepted: 24 November 2022
Published: 28 November 2022

1. Introduction

IgG4-related disease is a fibroinflammatory clinicopathological entity, characterized by dense infiltration of lymphocytes and IgG4-positive plasma cells in various tissues and organs [1,2]. The disease has been described in numerous organ systems, including pancreas, bile duct, lacrimal glands, salivary glands, central nervous system, thyroid, gastrointestinal tract, kidney, retroperitoneum, lymph nodes, and skin [2]. The specific involvement of ocular and orbital tissues is collectively termed IgG4-related ophthalmic disease (IgG4-ROD) [3]. IgG4-ROD can manifest in lacrimal glands, orbital soft tissue, extraocular muscles, trigeminal nerve branches, and the supraorbital or infraorbital nerves [3,4]. IgG4-ROD may involve a single structure or multiple structures within the orbit, and may also exhibit bilateral disease [4]. The first-line treatment for active disease is systemic glucocorticoids. Immunosuppressive drugs and biologic agents, especially rituximab, have also been proven effective [4–6]. However, after initial steroid treatment, high recurrence rates (up to 61%) have been reported in the literature [7–9]. Therefore, certain patients may benefit from maintenance therapy [5]. Radiotherapy has also been used as additional therapy for recurrent or refractory IgG4-ROD [6,10].

IgG4-ROD is usually associated with an increase in serum IgG4 levels. The diagnostic criteria for IgG4-ROD includes imaging studies and histopathologic examinations of ophthalmic tissues, and blood tests showing elevated serum IgG4 (≥135 mg/dL), requiring at least two out of three criteria for diagnosis [3]. In addition to contributing to the diagnosis [11,12], serum IgG4 levels have been useful in determining disease activity and predicting relapse [13–15]. After steroid treatment, patients with IgG4-ROD responded well in the early phase with reduction in IgG4 levels [13–15]. Elevated serum IgG4 levels before [7] and after [15] steroid treatment have both been reported as risk factors for IgG4-ROD relapse. Despite these findings, the long-term serial changes of serum IgG4 levels are not well understood. The aim of this study was to determine the association between long-term changes in serum IgG4 levels and the clinical course of IgG4-ROD.

2. Materials and Methods

This research was designed as a single-institution, retrospective case series. Data collection and chart review were approved by the Institutional Review Board of Taipei Veterans General Hospital, Taipei, Taiwan (IRB-TPEVGH No.2021-10-002CC). The study was conducted in accordance with the tenets of the Declaration of Helsinki.

Records of patients with diagnosis of IgG4-ROD from April 2010 to April 2020 were examined. We applied the diagnostic criteria for IgG4-ROD by Goto et al. [3], as follows: (1) imaging studies showing enlargement of the lacrimal gland, trigeminal nerve, or extraocular muscle as well as masses, enlargement, or hypertrophic lesions in various ophthalmic tissues; (2) histopathologic examination showing marked lymphocyte and plasmacyte infiltration, and sometimes fibrosis; IgG4+ plasmacytes found and the ratio of IgG4+ cells to IgG+ cells of 40% or above, or more than 50 IgG4+ cells per high-power field; (3) blood test showing elevated serum IgG4 (≥135 mg/dL). Diagnosis was classified as "definitive" when (1), (2), and (3) were satisfied; "probable" when (1) and (2) were satisfied; and "possible" when (1) and (3) were satisfied. All study participants met inclusion criteria of the following: (a) diagnosis of definite, probable or possible IgG4-ROD, according to the aforementioned criteria [3]; (b) patients received treatment and had follow-up for at least 1 year. Subjects were excluded from the study if they had IgG4 lymphoma (n = 5), sclerosing orbital inflammation [16] (more aggressive clinical course, resulting in permanent visual loss, n = 1), follow-up less than 1 year (n = 5), or not fulfilling diagnosis of IgG4-ROD (only meeting one criterion, n = 4).

Anatomic locations of ocular lesions were evaluated both clinically and by reviewing orbital CT or MRI scans. Systemic involvement was determined by clinical manifestations and imaging studies by ultrasound, CT, MRI or PET/CT.

Collected data included demographics, laterality of ocular disease (unilateral or bilateral), systemic involvement of IgG4 disease, ocular structure of involvement (lacrimal gland, orbital soft tissue, extraocular muscle, infraorbital nerve, etc.), presence of recurrent disease, serological results (IgG level at baseline; IgG4 level at baseline, lowest level after initial steroid treatment, during recurrence, and last follow-up), duration of steroid treatment, and total follow-up time.

Recurrence was defined as symptom re-emergence or lesion enlargement after initial steroid treatment. Remission was defined as resolution of clinical symptoms and/or radiologic lesions. For the main outcome measures, patients were categorized into groups based on last follow-up condition: Remission with normalized IgG4 levels (group 1), remission with elevated IgG4 levels (group 2), and stable with disease. A cutoff value of 135 mg/dL was used. For additional analysis, the study population was alternatively grouped into those with unilateral or bilateral ocular involvement.

Nonparametric tests were used for the following comparisons because most measurement data were not normally distributed. Categorical variables were analyzed using Fisher's exact tests. Measurements were compared between groups 1 and 2, and unilateral and bilateral ocular involvement using Mann–Whitney *U* tests. Statistical significance was defined as $p \leq 0.05$ and all analyses were conducted using SPSS statistical software (version 22, SPSS Inc., Chicago, IL, USA).

3. Results

A total of 25 patients diagnosed with IgG4-ROD were included in this study. On the basis of the diagnostic criteria for IgG4-ROD, the diagnosis was definite in 16 (64%) patients and possible in 9 (36%) patients. The demographics of the study population are summarized in Table 1. The mean age at diagnosis was 60.68 years. There were 17 (68%) men and 8 (32%) women. Ocular involvement was bilateral in 56% of patients. Systemic association was observed in 32% of patients. Disease involvement was isolated to one ocular structure in 60% of patients, while 28% and 12% of patients were concurrently affected in two and three ocular structures, respectively (not shown in table). The most common ocular structure involved was the lacrimal gland (76%), followed by orbital soft tissue (36%), extraocular muscle (20%) and infraorbital nerve (20%). Regarding disease course, 11 (44%) patients experienced recurrence during the follow-up period. The study population had initial steroid treatment of 14.53 ± 25.11 months (range 1.57 to 110 months), and follow-up period of 57.72 ± 33.36 months. According to last follow-up status, 9 (36%) patients had normalized IgG4 levels, and 16 (64%) patients had elevated IgG4 levels.

Table 1. Baseline characteristics of patients with IgG4-related ophthalmic disease (n = 25).

Characteristic	Value
Age (mean, year)	60.68
Male gender, n (%)	17 (68%)
Bilateral, n (%)	14 (56%)
Systemic involvement, n (%)	8 (32%)
Ocular involvement	
Lacrimal gland, n (%)	19 (76%)
Orbital soft tissue, n (%)	9 (36%)
Extraocular muscle, n (%)	5 (20%)
Infraorbital nerve, n (%)	5 (20%)
Recurrence, n (%)	11 (44%)
Duration of steroid treatment (mean, month)	14.53
Follow-up time, month (mean, month)	57.72
Last follow-up status of IgG4 level *	
Normalized, n (%)	9 (36%)
Elevated, n (%)	16 (64%)

* For IgG4 levels at last follow-up visit, a cutoff value of 135 mg/dL was used to differentiate between normalized and elevated values.

Figure 1 shows the serial changes in serum IgG4 levels from initial presentation to 60-month follow-up. The mean IgG4 level at baseline was the highest, with a steep decline to the lowest point after starting steroid treatment. After tapering and, for some patients, subsequently terminating steroid treatment, there was a gradual elevation in IgG4 levels. Later, the IgG4 levels fluctuated, and did not return to normalized levels in some patients. Table 2 shows comparison of patients with normal and elevated IgG4 levels according to last follow-up status; patients with normalized IgG4 levels had better response to initial steroid treatment and attained a significantly lower IgG4 level after treatment ($p = 0.002$). Table 3 further shows serum IgG4 levels at specific time points throughout the follow-up period, with the highest levels at baseline and disease recurrence (795.44 ± 553.10 mg/dL and 780.14 ± 568.38 mg/dL, respectively), lowest after initial treatment (176.36 ± 141.08 mg/dL), and 415.84 ± 345.87 mg/dL on last follow-up. At last follow-up visit, there was a distinct difference in IgG4 levels in patients with remission (mean 326.25 mg/dL) and those with stable disease (mean 699.55 mg/dL). In addition, in the 11 patients with recurrence, 3 had recurrent disease during tapering of steroids; as for the remaining 8 patients, the duration from discontinuing steroid treatment to recurrence was 11.92 ± 8.58 months (not shown in table).

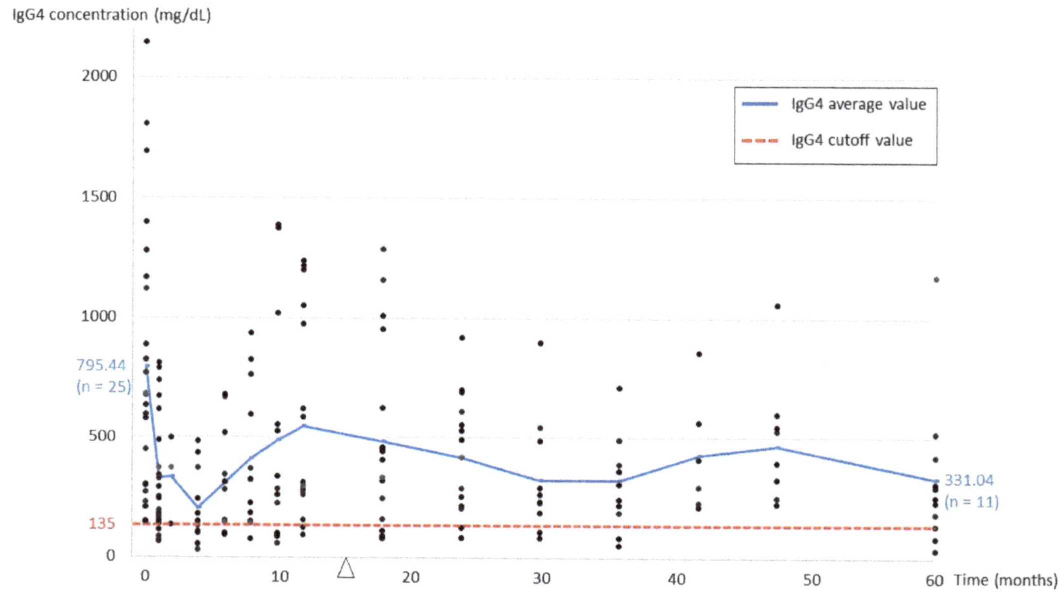

Figure 1. Serial changes in serum IgG4 levels from initial presentation to 60-month follow-up. The arrowhead indicates the end of initial steroid treatment (mean, 14.53 months).

Table 2. Comparison of patients with normal and elevated IgG4 levels according to last follow-up status.

Variable	IgG4 Normalized * (n = 9)	IgG4 Elevated * (n = 16)	*p* Value
Age (mean, year)	61.22	60.38	0.718
Male gender	77.8%	62.5%	0.661
Systemic involvement	44.4%	25%	0.394
Ocular involvement			
Lacrimal gland	55.6%	87.5%	0.142
Orbital soft tissue	44.4%	31.3%	0.671
Extraocular muscle	11.1%	25%	0.621
Infraorbital nerve	33.3%	12.5%	0.312
Recurrence	22.2%	56.3%	0.208
IgG level at baseline (mean, mg/dL)	1875.38	2227.13	0.636
IgG4 level at baseline (mean, mg/dL)	580.78	916.19	0.276
Lowest IgG4 level after initial steroid treatment (mean, mg/dL)	82.19	229.33	0.002
IgG4 level at last follow-up (mean, mg/dL)	78.50	605.60	0
Duration of steroid treatment (mean, month)	17.36	12.94	0.890

* For IgG4 levels at last follow-up visit, a cutoff value of 135 mg/dL was used to differentiate between normalized and elevated values.

After dividing the patients in remission into two groups depending on whether IgG4 levels were normalized (group 1) or elevated (group 2) on last follow up (Table 4), group 2 had a higher percentage of bilateral disease (70%), lacrimal gland involvement (90%) and recurrence (60%). In addition, throughout the disease course, group 2 had higher IgG4 levels at all time points documented: Baseline, lowest after initial treatment, and last follow up, albeit reaching statistical significance only at the lowest point (group 1, 82.19 ± 32.93 mg/dL vs. group 2, 152.80 ± 79.44 mg/dL, $p = 0.043$).

Table 3. Serial changes in serum IgG4 levels at specific time points.

Variable	Value
IgG level at baseline (mean, mg/dL) (n = 24, 1 missing datum)	2104.78
IgG4 level at baseline (mean, mg/dL) (n = 25)	795.44
IgG4/IgG ratio at baseline (n = 24, 1 missing datum)	0.3586
Lowest IgG4 level after initial steroid treatment (mean, mg/dL) (n = 25)	176.36
IgG4 level during recurrence (mean, mg/dL) (n = 11)	780.14
IgG4 level at baseline in patients with recurrence (mean, mg/dL) (n = 11)	789.44
IgG4 level at last follow-up (mean, mg/dL) (n = 25)	415.84
In patients with remission (mean, mg/dL) (n = 19)	326.25
In patients with stable disease (mean, mg/dL) (n = 6)	699.55

Table 4. Comparison of patients in remission with normal and elevated IgG4 levels.

Variable	Group 1 (Remission with Normalized IgG4) [1] (n = 9)	Group 2 (Remission with Elevated IgG4) [1] (n = 10)	*p* Value
Last follow-up IgG4 (mean, mg/dL)	78.50	549.23	NA [2]
Age (mean, year)	61.22	63.10	0.905
Male gender	77.80%	60%	0.628
Bilateral	33.30%	70%	0.179
Systemic involvement	44.40%	20%	0.35
Ocular involvement			
Lacrimal gland	55.56%	90%	0.141
Orbital soft tissue	44.44%	20%	0.35
Extraocular muscle	11.11%	30%	0.582
Infraorbital nerve	33.33%	20%	0.628
Recurrence	22.22%	60%	0.17
IgG level at baseline (mean, mg/dL)	1875.38	1857.67	0.888
IgG4 level at baseline (mean, mg/dL)	553.95	701.37	0.696
Lowest IgG4 level after initial steroid treatment (mean, mg/dL)	82.19	152.80	0.043
Reduction in IgG4 level (%) [3]	75.58%	75.68%	0.847
Duration of steroid treatment (mean, month)	17.36	5.84	0.968

[1] For IgG4 levels at last follow-up visit, a cutoff value of 135 mg/dL was used to differentiate between normalized and elevated values. [2] NA, not applicable. [3] Reduction in IgG4 level (%) = Absolute value of (Lowest level–Baseline level)/Baseline level × 100%.

The study population was alternatively grouped into those with unilateral or bilateral ocular involvement (Table 5). All patients with bilateral ocular disease had involvement of the lacrimal gland, whereas only 45.5% of patients with unilateral disease were involved ($p = 0.003$). Patients with bilateral disease also had higher IgG and IgG4 levels at baseline, and a greater reduction in IgG4 levels after initial steroid treatment ($p = 0.007$, 0.004 and 0.005, respectively).

Table 5. Comparison of parameters between patients with unilateral and bilateral ocular disease.

Variable	Unilateral (n = 11)	Bilateral (n = 14)	*p* Value
Age (mean, year)	63.00	58.86	0.291
Male gender	63.6%	71.4%	1
Systemic involvement	36.4%	28.6%	1
Ocular involvement			
Lacrimal gland	45.5%	100.0%	0.003
Orbital soft tissue	45.5%	28.6%	0.434
Extraocular muscle	18.2%	21.4%	1
Infraorbital nerve	9.1%	28.6%	0.341

Table 5. *Cont.*

Variable	Unilateral (n = 11)	Bilateral (n = 14)	*p* Value
Recurrence	44.4%	40.0%	1
IgG level at baseline (mean, mg/dL)	1649.55	2522.08	0.007
IgG4 level at baseline (mean, mg/dL)	455.09	1083.43	0.004
Lowest IgG4 level after initial steroid treatment (mean, mg/dL)	159.49	189.61	0.572
IgG4 during recurrence (mean, mg/dL)	205.80	1186.00	0.114
IgG4 level at last follow-up (mean, mg/dL)	281.98	521.02	0.183
Reduction in IgG4 level (%) [1]	64.72	81.50	0.005
Change in IgG4 level at recurrence (%) [2]	−8.21	21.93	0.667
Duration of steroid treatment (mean, month)	10.83	17.44	0.066

[1] Reduction in IgG4 level (%) = Absolute value of (Lowest level−Baseline level)/Baseline level × 100%. [2] Change in IgG4 level at recurrence (%) = (Recurrence level−Baseline level)/Baseline level × 100%.

4. Discussion

We present a long-term follow-up of patients with IgG4-ROD focusing on the serial changes of serum IgG4 levels. To our knowledge, this case series of IgG4-ROD patients had the longest follow-up period compared with previous studies. In this cohort, up to 64% of patients had elevated IgG4 levels at last follow-up, which led us to investigate the correlation between IgG4 levels and their clinical significance.

Out of all patients in remission, the IgG4 levels failed to normalize in 10 of 19 patients (52.6%). The patients with elevated IgG4 concentrations had a relatively higher rate of bilateral disease, lacrimal gland involvement and recurrence. These patients also had higher IgG4 titers throughout the follow-up period, with statistically higher titers compared to the normalized group at the lowest point after initial treatment. These clinical and serological factors may indicate greater disease activity and poorer response to initial steroid treatment, rendering this group of patients prone to a higher IgG4 titer in the long term. Similarly, Wallace et al. [4] found the serum IgG4 level among those with bilateral disease to be higher compared to those with unilateral disease. Interestingly, serum IgG4 was higher in patients with lacrimal or salivary gland lesions compared to those with other affected organs [17,18], but no previous studies had compared serum IgG4 levels amongst different ocular structures involved, which is perhaps difficult because multiple ocular lesions can coexist. In a smaller case series of nine patients with IgG4-ROD and shorter follow-up time (average of 8.6 months), six patients had IgG4 levels at last follow-up, where two patients had clinical relapse, one patient had a normalized IgG4 level, and three patients had elevated IgG4 levels [14], but the case number may be too limited to come to comparable conclusions. A multicenter study on autoimmune pancreatitis showed that IgG4 levels failed to normalize in 63% of patients after treatment with oral steroids, but only 30% of these patients with persistent elevation of serum IgG4 levels had relapses [19]. We infer that IgG4-ROD patients with better response to initial steroid treatment would be more likely to resume normalized IgG4 levels in the long-term. In addition, despite persistent elevations of serum IgG4 levels, some patients can still remain in remission.

IgG4 is one of four subclasses of IgG, typically less than 5% of total IgG [20]. In general, IgG4 is a benign, non-pathogenic antibody. From a molecular structural standpoint, the disrupted C1q binding site of IgG4 contributes to its inability to activate the complement cascade [21]. The low affinity for classical Fcγ-receptors makes IgG4 inefficient in activating effector systems [21]. In addition, IgG4 forms small, non-precipitating immune complexes due to effective monovalency [22]. The pathophysiology of IgG4-related disease consists of multiple immune-mediated mechanisms, which remains to be fully elucidated. The counter-inflammatory properties of IgG4 and the fact that disease-specific IgG4 autoantibodies have not been found suggest that serum IgG4 is a response to inflammatory stimulus, more likely a consequence than a cause of the disease [23]. Two possible explanations for elevated IgG4 levels in patients in remission are "blocking antibody" and indicator of tolerance induction. IgG4 tends to appear after prolonged immunization with a decrease in symptoms. This is thought to be due to competition with IgE for allergen by IgG4, i.e., the blocking antibody, and inhibiting mast cell degranulation and/or preventing IgE-facilitated activation of T cells, and a suppression of the late-phase reaction [24,25]. Furthermore, IgG4 is a marker of tolerance induction. Usually at a single time point, the IgG4 levels in symptomatic patients are higher compared to those in asymptomatic patients. In long-term follow-up, a considerable increase in IgG4 can indicate that anti-inflammatory, tolerance-inducing mechanisms have been activated, in which inappropriate inflammatory reactions are reduced [20]. Therefore, regardless of IgG4 levels in patients with remission, we believe they can have regular follow up without intervention if they remain asymptomatic.

Approximately half of the study population had bilateral ocular lesions. These patients had statistically higher baseline IgG and baseline IgG4 levels compared to unilateral disease. This is in agreement with previous studies, which showed bilateral disease accounting for 49–62% of patients [26,27]. Wallace et al. [4] also found the baseline serum IgG4 level among those with bilateral disease to be statistically higher than unilateral disease. The reason for this finding is likely due to a higher disease activity in bilateral involvement. In addition, patients with bilateral disease also had better response to initial steroid therapy, resulting in a significantly greater reduction in IgG4 level. Interestingly, the change in serum IgG4 levels at recurrence (compared to baseline) differed between the two groups, with unilateral disease seeing a decrease in IgG4 values, and bilateral disease with an increase from baseline, albeit not reaching statistical significance. We can infer that there is a difference in the clinical course of unilateral and bilateral ocular lesions.

The lacrimal gland was the most common orbital structure affected in IgG4-ROD in our series and many previous studies [4,9,26,28,29]. Overall, the most frequently involved structure was the lacrimal gland (76%), followed by orbital soft tissue (36%), extraocular muscle (20%) and infraorbital nerve (20%). Although the percentage of ocular lesions varied with each study [4,9,26,28,29], our results were largely consistent with the results of a multicenter study by Goto et al. [28], which showed pathologic lesions in the lacrimal glands (86%), isolated and diffuse orbital lesions (19%), extraocular muscles (21%) and trigeminal nerve (20%). The high percentage of lacrimal gland involvement in IgG4-ROD may also explain why all patients with bilateral disease had lacrimal gland lesions in our study.

A few studies have looked into the treatment outcomes of IgG4-ROD. Regarding treatment regimens, an initial combination of glucocorticoid and immunosuppressant therapy had advantages over glucocorticoid monotherapy, with a longer relapse-free survival time [30], a lower relapse rate and a shorter glucocorticoid therapy duration [9]. Risk factors for relapse include multiple ocular lesions [9], extraocular muscle and/or trigeminal nerve enlargements [8], and additional radiotherapy after surgical debulking with oral steroids [27]. In the present series, with a mean follow-up duration of 57.72 months, 19 (76%) patients were in remission while 6 (24%) patients had stable disease. We infer that patients with IgG4-ROD have a favorable outcome, regardless of their serologic status in the long-term.

There are some limitations of the current study that should be addressed. First, the study is a retrospective, non-randomized controlled investigation, which might limit the generalizability of our findings. Second, nine patients (36%) did not receive biopsy and were diagnosed as "possible" IgG4-ROD. This is mainly due to difficulty in obtaining specimens in patients with extraocular muscle or deep orbital soft tissue involvement. Nevertheless, the findings of the present study have important implications for understanding the prognosis of IgG4-ROD.

5. Conclusions

Our results demonstrated that IgG4-ROD patients with greater response to initial steroid therapy, manifesting with lower serum IgG4 levels, would be more inclined to have normalized IgG4 levels in the long term. In addition, some patients can remain in remission despite persistently elevated serum IgG4 levels, and they can be followed up without treatment unless disease relapse. Moreover, different baseline serum IgG4 levels and disparities in disease course were observed between unilateral and bilateral ocular lesions. Future studies are needed to provide further insight into this complex disease.

Author Contributions: Conceptualization, W.-Y.C. and C.-C.T.; methodology, W.-Y.C. and C.-C.T.; validation, W.-Y.C. and C.-C.T.; formal analysis, W.-Y.C. and C.-C.T.; investigation, W.-Y.C. and C.-C.T.; resources, C.-C.T.; writing, W.-Y.C.; validation, C.-Y.T.; supervision, C.-C.T. and C.-Y.T. All authors have read and agreed to the published version of the manuscript.

Funding: This study was funded by the Ministry of Science and Technology, Executive Yuan Taiwan, grant numbers MOST 110-2314-B-075-064-MY2 and Taipei Veterans General Hospital, Taiwan V111-C-083.

Institutional Review Board Statement: The current study was in compliance with recognized international standards and the principles of the Declaration of Helsinki in 1964 and its late amendments. Moreover, our study was approved by the Institutional Review Board of Taipei Veterans General Hospital (IRB-TPEVGH No.2021-10-002CC).

Informed Consent Statement: Patient consent was waived by the Institutional Review Board of Taipei Veterans General Hospital due to the retrospective nature.

Data Availability Statement: Not applicable.

Conflicts of Interest: The authors declare no conflict of interest.

References

1. Kamisawa, T.; Funata, N.; Hayashi, Y.; Eishi, Y.; Koike, M.; Tsuruta, K.; Okamoto, A.; Egawa, N.; Nakajima, H. A new clinicopathological entity of IgG4-related autoimmune disease. *J. Gastroenterol.* **2003**, *38*, 982–984. [CrossRef] [PubMed]
2. Umehara, H.; Okazaki, K.; Masaki, Y.; Kawano, M.; Yamamoto, M.; Saeki, T.; Matsui, S.; Yoshino, T.; Nakamura, S.; Kawa, S.; et al. Comprehensive diagnostic criteria for IgG4-related disease (IgG4-RD), 2011. *Mod. Rheumatol.* **2012**, *22*, 21–30. [CrossRef] [PubMed]
3. Goto, H.; Takahira, M.; Azumi, A. Japanese Study Group for IgG4-Related Ophthalmic Disease. Diagnostic criteria for IgG4-related ophthalmic disease. *Jpn. J. Ophthalmol.* **2015**, *59*, 1–7. [CrossRef] [PubMed]
4. Wallace, Z.S.; Deshpande, V.; Stone, J.H. Ophthalmic manifestations of IgG4-related disease: Single-center experience and literature review. *Semin. Arthritis Rheum.* **2014**, *43*, 806–817. [CrossRef]
5. Khosroshahi, A.; Wallace, Z.S.; Crowe, J.L.; Akamizu, T.; Azumi, A.; Carruthers, M.N.; Chari, S.T.; Della-Torre, E.; Frulloni, L.; Goto, H.; et al. International Consensus Guidance Statement on the Management and Treatment of IgG4-Related Disease. *Arthritis Rheumatol.* **2015**, *67*, 1688–1699. [CrossRef] [PubMed]
6. Detiger, S.E.; Karim, A.F.; Verdijk, R.M.; Van Hagen, P.M.; Van Laar, J.A.M.; Paridaens, D. The treatment outcomes in IgG4-related orbital disease: A systematic review of the literature. *Acta Ophthalmol.* **2019**, *97*, 451–459. [CrossRef] [PubMed]
7. Hong, J.W.; Kang, S.; Song, M.K.; Ahn, C.J.; Sa, H. Clinicoserological factors associated with response to steroid treatment and recurrence in patients with IgG4-related ophthalmic disease. *Br. J. Ophthalmol.* **2018**, *102*, 1591–1595. [CrossRef]
8. Kubota, T.; Katayama, M.; Nishimura, R.; Moritani, S. Long- term outcomes of ocular adnexal lesions in IgG4- related ophthalmic disease. *Br. J. Ophthalmol.* **2020**, *104*, 345–349. [CrossRef] [PubMed]
9. Gan, L.; Luo, X.; Fei, Y.; Peng, L.; Zhou, J.; Li, J.; Lu, H.; Liu, Z.; Zhang, P.; Liu, X.; et al. Long-Term Outcomes of IgG4-Related Ophthalmic Disease in a Chinese IgG4-Related Disease Cohort. *Front. Med.* **2021**, *8*, 784520. [CrossRef]
10. Lin, Y.H.; Yen, S.H.; Tsai, C.C.; Kao, S.C.; Lee, F.L. Adjunctive Orbital Radiotherapy for Ocular Adnexal IgG4-related Disease: Preliminary Experience in Patients Refractory or Intolerant to Corticosteroid Therapy. *Ocul. Immunol. Inflamm.* **2015**, *23*, 162–167. [CrossRef]
11. Xu, W.L.; Ling, Y.C.; Wang, Z.K.; Deng, F. Diagnostic performance of serum IgG4 level for IgG4-related disease: A meta-analysis. *Sci. Rep.* **2016**, *6*, 32035. [CrossRef] [PubMed]
12. Xia, C.S.; Fan, C.H.; Liu, Y.Y. Diagnostic performances of serum IgG4 concentration and IgG4/IgG ratio in IgG4-related disease. *Clin. Rheumatol.* **2017**, *36*, 2769–2774. [CrossRef] [PubMed]
13. Yu, W.K.; Kao, S.C.; Yang, C.F.; Lee, F.L.; Tsai, C.C. Ocular adnexal IgG4-related disease: Clinical features, outcome, and factors associated with response to systemic steroids. *Jpn. J. Ophthalmol.* **2015**, *59*, 8–13. [CrossRef] [PubMed]
14. Hagiya, C.; Tsuboi, H.; Yokosawa, M.; Hagiwara, S.; Hirota, T.; Takai, C.; Asashima, H.; Miki, H.; Umeda, N.; Horikoshi, M.; et al. Clinicopathological features of IgG4-related disease complicated with orbital involvement. *Mod. Rheumatol.* **2014**, *24*, 471–476. [CrossRef]
15. Woo, Y.J.; Kim, J.W.; Yoon, J.S. Clinical implications of serum IgG4 levels in patients with IgG4-related ophthalmic disease. *Br. J. Ophthalmol.* **2017**, *101*, 256–260.
16. Son, K.Y.; Woo, K.I.; Kim, Y.D. Clinical Outcomes of IgG4-Related Ophthalmic Disease and Idiopathic Sclerosing Orbital Inflammation. *Ophthalmic Plast. Reconstr. Surg.* **2022**, *38*, 34–39. [CrossRef]

17. Zen, Y.; Nakanuma, Y. IgG4-related disease: A cross-sectional study of 114 cases. *Am. J. Surg. Pathol.* **2010**, *34*, 1812–1819. [CrossRef]
18. Hamano, H.; Arakura, N.; Muraki, T.; Ozaki, Y.; Kiyosawa, K.; Kawa, S. Prevalence and distribution of extrapancreatic lesions complicating autoimmune pancreatitis. *J. Gastroenterol.* **2006**, *41*, 1197–1205. [CrossRef]
19. Kamisawa, T.; Shimosegawa, T.; Okazaki, K.; Nishino, T.; Watanabe, H.; Kanno, A.; Okumura, F.; Nishikawa, T.; Kobayashi, K.; Ichiya, T.; et al. Standard steroid treatment for autoimmune pancreatitis. *Gut* **2009**, *58*, 1504–1507. [CrossRef]
20. Aalberse, R.C.; Stapel, S.O.; Schuurman, J.; Rispens, T. Immunoglobulin G4: An odd antibody. *Clin. Exp. Allergy* **2009**, *39*, 469–477. [CrossRef]
21. Davies, A.M.; Sutton, B.J. Human IgG4: A structural perspective. *Immunol. Rev.* **2015**, *268*, 139–159. [CrossRef] [PubMed]
22. Van Der Zee, J.S.; Van Swieten, P.; Aalberse, R.C. Serologic aspects of IgG4 antibodies. II. IgG4 antibodies form small, nonprecipitating immune complexes due to functional monovalency. *J. Immunol.* **1986**, *137*, 3566–3571. [PubMed]
23. Stone, J.H.; Zen, Y.; Deshpande, V. IgG4-related disease. *N. Engl. J. Med.* **2012**, *366*, 539–551. [CrossRef] [PubMed]
24. Dodev, T.S.; Bowen, H.; Shamji, M.H.; Bax, H.J.; Beavil, A.J.; McDonnell, J.M.; Durham, S.R.; Sutton, B.J.; Gould, H.J.; James, L.K. Inhibition of allergen-dependent IgE activity by antibodies of the same specificity but different class. *Allergy* **2015**, *70*, 720–724. [CrossRef] [PubMed]
25. Santos, A.F.; James, L.K.; Bahnson, H.T.; Shamji, M.H.; Couto-Francisco, N.C.; Islam, S.; Houghton, S.; Clark, A.T.; Stephens, A.; Turcanu, V.; et al. IgG4 inhibits peanut-induced basophil and mast cell activation in peanut-tolerant children sensitized to peanut major allergens. *J. Allergy Clin. Immunol.* **2015**, *135*, 1249–1256. [CrossRef] [PubMed]
26. Yu, W.K.; Tsai, C.C.; Kao, S.C.; Liu, C.J.L. Immunoglobulin G4-related ophthalmic disease. *Taiwan J. Ophthalmol.* **2018**, *8*, 9–14. [CrossRef] [PubMed]
27. Chen, J.; Zhang, P.; Ye, H.; Xiao, W.; Chen, R.; Mao, Y.; Ai, S.; Liu, Z.; Tang, L.; Yang, H. Clinical features and outcomes of IgG4-related idiopathic orbital inflammatory disease: From a large southern China-based cohort. *Eye* **2021**, *35*, 1248–1255. [CrossRef]
28. Goto, H.; Ueda, S.I.; Nemoto, R.; Ohshima, K.I.; Sogabe, Y.; Kitagawa, K.; Ogawa, Y.; Oyama, T.; Furuta, M.; Azumi, A.; et al. Clinical features and symptoms of IgG4-related ophthalmic disease: A multicenter study. *Jpn. J. Ophthalmol.* **2021**, *65*, 651–656. [CrossRef]
29. Andrew, N.; Kearney, D.; Selva, D. IgG4-related orbital disease: A meta-analysis and review. *Acta Ophthalmol.* **2013**, *91*, 694–700. [CrossRef]
30. Zhao, Z.; Mou, D.; Wang, Z.; Zeng, Q.; Wang, Z.; Xue, J.; Ren, L.; Liu, Y.; Su, Y. Clinical features and relapse risks of IgG4-related ophthalmic disease: A single-center experience in China. *Arthritis Res. Ther.* **2021**, *23*, 98. [CrossRef]

Journal of
Personalized Medicine

Article

Morphological and Functional Aspects and Quality of Life in Patients with Achromatopsia

Caroline Chan, Berthold Seitz and Barbara Käsmann-Kellner *

Department of Ophthalmology, University of Saarland Medical Center in Homburg/Saar, 66421 Homburg/Saar, Germany; caroline.chan@uks.eu (C.C.); berthold.seitz@uks.eu (B.S.)
* Correspondence: kaesmann@gmail.com

Abstract: (1) Background: Achromatopsia is a rare disease of which the natural course and impact on life are still unknown to this date. We aimed to assess the morphological, functional characteristics, and quality of life in a large sample size of patients with achromatopsia. (2) A total of 94 achromats were included in this retrospective cohort study. Sixty-four were patients of the Department of Ophthalmology, Saarland University Medical Centre in Homburg/Saar, Germany, between 2008 and 2021. Thirty further participants with achromatopsia from the national support group were included using an online questionnaire, which is available under 'Supplementary data'. Statistical analysis was performed using SPSS Version 25; (3) The 94 patients (37 males (39.4%) and 57 females (60.6%)) showed a mean age of 24.23 ± 18.53 years. Visual acuity was stable (SD ± 0.22 logMAR at 1.0 logMAR) over a time of observation from 2008 to 2021. Edge filter glasses were the most used optical aids, while enlarged reading glasses were the most used low vision aids. (4) Conclusions: Our findings give an insight into describing the natural process and the quality of life of achromatopsia. The results demonstrate that achromatopsia is a predominantly stationary disease. The individual prescription of edge filters and low-vision aids is essential following a personalised fitting.

Keywords: achromatopsia; cone; low vision aids; quality of life; personalised ophthalmology

Citation: Chan, C.; Seitz, B.; Käsmann-Kellner, B. Morphological and Functional Aspects and Quality of Life in Patients with Achromatopsia. *J. Pers. Med.* **2023**, *13*, 1106. https://doi.org/10.3390/jpm13071106

Academic Editor: Chieh-Chih Tsai

Received: 15 June 2023
Revised: 2 July 2023
Accepted: 5 July 2023
Published: 7 July 2023

1. Introduction

Our study's scope was to analyse the morphological changes that may be associated with achromatopsia, to follow these changes in visual acuity, and to assess the impact of achromatopsia on patients' quality of life. As far as we know, to this date, the nature of achromatopsia has still not been clearly defined. Furthermore, no study on achromatopsia has been conducted with a large sample size focusing on the psychosocial aspect. Achromatopsia, also called rod monochromatism or total colour blindness, belongs to cone dysfunction syndromes and is present at birth [1]. It is considered a rare disease with a prevalence of 1/30,000 births worldwide [2]. It is an inherited, autosomal recessive disease, with typical characteristics such as a congenital reduction in visual acuity to around 20/200, photophobia, lack of colour discrimination, and nystagmus. The fundus examination usually does not show pathological changes, but sometimes there can be a constriction of blood vessels and pathological macular findings in the adult age [3–5]. Mutations of the six genes *CNGA3* [6], *CNGB3* [7], *GNAT2* [8], *PDE6H* [9], *PDE6C* [10], which play a vital role in the phototransduction cascade and *ATF6* [11], which regulates the unfolded protein response, have been identified in achromatopsia.

Refractive errors are common [12]. There are two types of achromatopsia: In 'complete' achromatopsia, all three types of cones (1. Blue, 2. Green, 3. Red) do not function, while in 'incomplete' achromatopsia, there may be the partial function of one or more cone types [13].

Up to this date, achromatopsia treatment has mainly focused on managing the symptoms with different optical and electronic low-vision aids [3,14]. The first genetic treatment studies for *CNGA3* and *CNGB3* are underway; however, the results in adults have not shown vigorous improvements. While there is currently no known cure, many management aids can alleviate some symptoms associated with ACHM. In particular, photoaversion has been described to be one of the more debilitating symptoms of achromatopsia, with the need for different aids [15]. (Figure 1) These include red-tinted contact lenses, sunglasses, wrap-around glasses, and dark and red-filtered glasses, which have been shown to reduce photophobia and improve visual acuity. These optical aids have to be tried and fitted individually for the best benefit of the patient. Especially red-tinted contact lenses have been a relief for achromats, as they are more discrete than dark and red-filtered glasses. (Figure 2) Low vision aids such as high-powered magnifiers and electronic devices can be useful for reading. Children with ACHM should be given preferential seating in class to benefit fully from these aids (such as at the front and to avoid the impact of glare on vision or away from windows) [2,14].

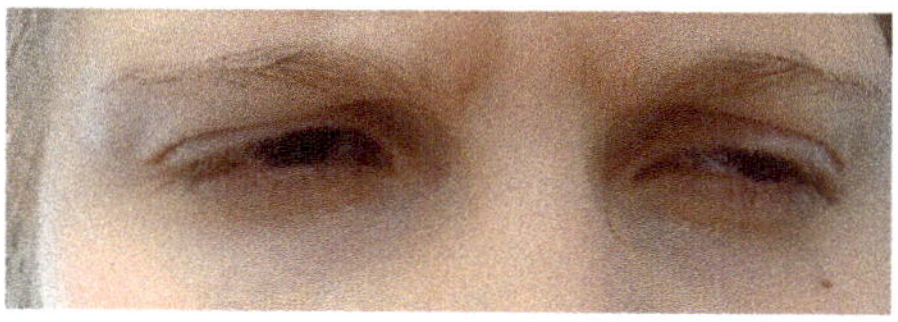

Figure 1. A 15-year-old girl without edge filter, not making eye contact.

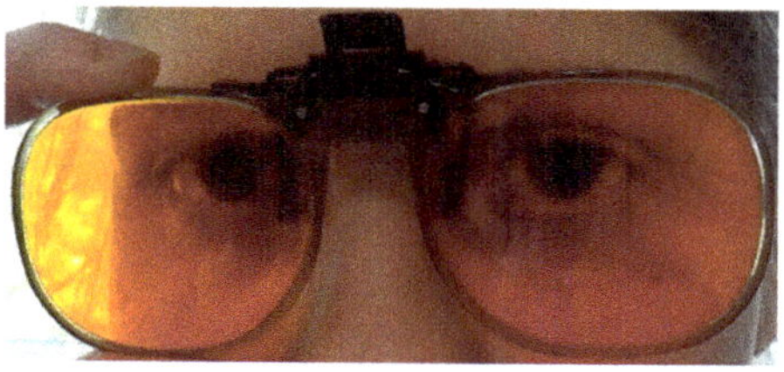

Figure 2. Same girl with edge filter glasses, eyes opened normally and eye contact [16].

It has recently been shown that gene therapy using an adeno-associated virus could perhaps provide a promising treatment for achromatopsia. Recently, human clinical trials have been developed, which primarily showed the safety and efficacy of the treatment [17–19]. There is still an ongoing debate concerning the natural course of achromatopsia. Some studies report that it is a progressive disorder, suggesting that gene therapy should be applied at an earlier age [20–24]. By contrast, other studies imply that it is a predominantly stationary condition, indicating that there might be a larger window for gene therapy [25–28]. However, these studies had small group sizes, with Hirji et al. having the largest with 50 subjects [28]. Furthermore, research has rarely been performed on the quality of life and subjective impressions of patients with achromatopsia. Very few studies have evaluated the influence of complete colour vision loss on the quality of life of these people [29]. The only study that came close to quality of life was Aboshiha et al.'s, which investigated the impact of photophobia on the life of achromats [15]. With human gene therapy trials developing rapidly, it is fundamental to elucidate the natural process of achromatopsia and find correlations between the genotype and phenotype to choose suitable candidates and the right time for treatment. In addition, understanding how achromatopsia affects the lives of these patients could help in identifying strategies that can improve patient outcomes.

2. Materials and Methods

2.1. Participants

In this retrospective cohort study, a total of 94 participants with a confirmed diagnosis of achromatopsia were included. The diagnosis was established by clinical presentation (severe glare sensitivity with normal ocular findings, in several cases with electrophysiological confirmation and with genetic evaluation). Exclusion criteria included atypical clinical findings (no glare sensitivity), no acceptance of edge filters, morphologically visible pathological findings (such as in albinism which presents with glare sensitivity as well), and negative genetic testing in patients who were genetically tested. Sixty-four of the participants were patients of the Department of Ophthalmology, Saarland University Medical Centre in Homburg/Saar, Germany, between 2008 and 2021. Thirty additional participants who completed a standardized questionnaire were members of the national support association of achromatopsia. The age range was very broad as we aimed to include all patients with this rare disease. The research was approved by the local ethics committee of Saarland (Nr. 85-21) and followed the principles of the Declaration of Helsinki. Informed consent was provided by all subjects.

2.2. Medical Records

Data from the patient's medical history were collected using the clinic's electronic charts. Patients underwent a complete ophthalmic examination: Visual acuity (Landolt eye chart), colour vision (Ishihara and Panel D 15 test), refraction (retinoscopy), anterior segment findings (slit-lamp), central and peripheral retina and optic nerve head (ophthalmoscope), and strabismus (prism cover test).

2.3. Questionnaire

A survey was conducted from April to May 2021 to analyse the quality of life that people living with achromatopsia had. This survey was distributed to all of the patients in our hospital diagnosed with achromatopsia and to the members of the national self-help group of achromatopsia (Achromatopsie Selbsthilfe e.V. Dorsten, Germany, www.achromatopsie.de, accessed on 20 March 2021). This survey consisted of 18 main questions comprising demographic and non-demographic questions based on their quality of life, such as education and visual aids (Table 1).

Table 1. The 18 main questions asked in our questionnaire to individuals with achromatopsia.

(1). Are you a member in a visually impaired association?
(2). What is your age and gender?
(3). Where was your ophthalmologic treatment?
(4). Was your diagnosis of achromatopsia genetically confirmed?
(5). When was the diagnosis of achromatopsia made?
(6). What is your visual acuity?
(7). What is/was your school education?
(8). What is/was your vocational training?
(9). What is/was your profession?
(10). Do you own a disability ID card?
(11). Do you receive any blind allowances?
(12). Which optical aids/light protection do you use?
(13). Which magnifying aids do you use?
(14). Which other aids do you use?
(15). Which computer programs/apps do you use?
(16). Do you use a colour naming system?
(17). What is the visual issue you suffer most from?
(18). How is your personal experience living with achromatopsia?

Most demographic questions were multiple-choice with a few open-ended questions. A free-text box was added after each question to let participants share any additional thoughts or information. The following criteria were retrieved from their medical history, and the questionnaire was taken into consideration: gender, age, age of diagnosis, past visual acuity, actual visual acuity, standardised colour test, subjective colour perception, refraction, morphological findings, strabismus, academic career, practiced occupation, present low vision aids, degree of handicap, allowances for the blind, mutated gene and primary concern of the patients.

2.4. Statistical Analysis

The software SPSS Statistics, IBM, Version 25, was used to enter and evaluate our data. Continuous data were described as the mean, standard deviation, median, minimum, and maximum. Categorical variables were described as percentages and were compared using the Chi-Square test. The Wilcoxon test was used for non-parametric variables. Spearman tests determined the relationship between non-parametric variables. Spearman's ρ (rho) is a rank correlation coefficient that was used for two ordinal variables or when one variable had a continuous normal distribution; the other variable was categorical or non-normally distributed. This test was a non-parametric test, which could be used with variables that had a non-normal distribution. p values < 0.05 were considered statistically significant.

3. Results

We took two approaches to collect the data: patients' records collected from our clinic's database and the questionnaire we created for the purpose of this study. These clinical records were used to evaluate morphological and functional characteristics, while the questionnaire assessed the functional aspects and the quality of life of achromats. Out of the 48 clinical examinations performed in domo, three of the patients did not request a genetic analysis (Table 2).

Table 2. Method of data collection for the population of participants.

	Total Number out of 94 Achromats	**Percentage Number out of 94 Achromats**
Only clinical examinations performed in domo	48	51%
Out of clinical examinations performed in domo; Molecular genetics confirmed	45	48%
Clinical examinations performed in domo + answered questionnaire	16	17%
Only answered questionnaire	30	32%

3.1. Demographics and Genes

Ninety-four participants were included in this study, with 37 males (39,4%) and 57 females (60.6%). The mean age was 24.23 years (SD ± 18.53 years). In total, 32 (34%) subjects presented with mutations in the *CNGB3* gene, 12 (12.8%) subjects had a *CNGA3* mutation, one (1.1%) subject had the *PDE6C* gene, and for 49 subjects (52.1%), no genetic analysis had been made up to now, mostly because the patients saw no necessity in spite of being informed about ongoing clinical trials.

3.2. Functional Findings

To confirm that achromatopsia is a stationary disease (functional aspect), we analysed the changes in visual acuity, as well as the refractive/orthoptic changes, with the increase in age in the achromats.

3.3. Visual Acuity

Out of the 62 patients examined ophthalmologically, the first known average visual acuity was 1.0 logMAR (SD ± 0.22), ranging between 0.42 logMAR and 1.52 logMAR. Out of 91 patients, including those answering the questionnaire, the actual average visual acuity was also 1.0 logMAR (SD ± 0.23), ranging from 0.22 logMAR to 1.70 logMAR.

We evaluated the difference between the visual acuity during the first and final visit to our clinic with an observation time of 1 to 21 years. The results indicated no significant difference between both measurements (r = 0.239, p = 0.811).

Out of 91 patients, no correlation was found between visual acuity and age (r = 0.002 and p = 0.985) (Figure 3). This confirmed the assumption that achromatopsia is a functionally stable disease over the years.

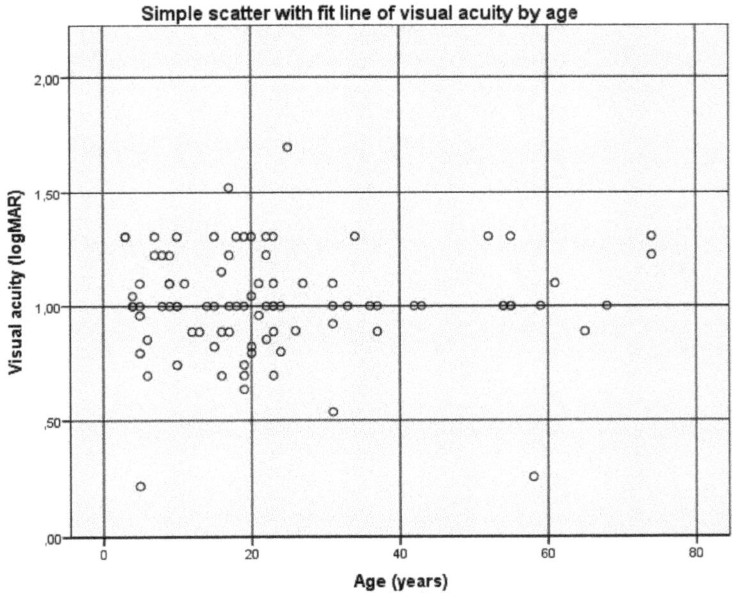

Figure 3. There was no significant decrease in visual acuity as a function of age. Visual acuity is expressed in logMAR and age in years. No correlation was found between visual acuity and age (r = 0.002 and p = 0.985).

Out of the 62 patients seen and examined ophthalmologically, 12 had the *CNGB3* gene, 32 had the *CNGB2* gene, 1 had the *PDE6C* gene, and 17 did not have a genetic analysis. We analysed if any of the six genes responsible for achromatopsia were linked to visual impairment. Out of 45 patients who received a genetic diagnosis, no correlation was found between the type of gene and visual acuity (r = −0.158, p = 0.311).

3.4. Refractive and Orthoptic Findings (Refraction, Nystagmus, Strabismus)

Out of 62 patients, the spherical refraction mean was 1.83 diopters (SD ± 3.76, −4.38 to +8.5 dpt). No correlation was found between the mutated genes and the degree of spherical refraction (r = −0.075, 0.669). The mean astigmatic refraction was −1.86 diopters (SD ± 0.88, −3.0 to −0.5 dpt). No correlation between the visual acuity and spherical refraction was found (r = −0.111, p = 0.393), between the visual acuity and astigmatic refraction (r = 0.144, p = 0.380), and between the mutated genes and the astigmatic refraction (r = 0.097, p = 0.578). Out of 61 patients, 54 (88.5%) had nystagmus, while 7 (11.5%) did not present with nystagmus. No statistical tests were used to analyse the frequency of nystagmus. As many conatal visually handicapped people were present with strabismus, we tested these for achromatopsia as well. Out of 58 patients, 34 (58.6%) did not present any strabismus, 6 (10.3%) presented a strabismus convergens, 17 (29.3%) had a strabismus divergens, and 1 (1.7%) presented a strabismus convergens and verticalis. No correlation was found between age and strabismus (r = 0.057, p = 0.672). No correlation was found between the mutated gene and strabismus (r = −0.230, p = 0.213).

3.5. Morphological Findings

We analysed the relationship between the genotype and phenotype, as well as the age and phenotype. The phenotype consisted of changes in the anterior segments, cataracts, retinal arterial vessels, the optic nerve, the peripheral retina, the fundus periphery, and the macula. Out of 62 subjects, a majority of 59 (95.2%) patients did not have any pathological changes in the anterior segment findings, with 3 (4.8%) over 50 years old presenting with a cataract, which was interpreted as an age-related finding and was not associated with achromatopsia. As the three patients with an incipient cataract had no significantly different visual acuity, we did not exclude them from the patient group. Out of 60 subjects, 32 (53.3%) had regular retinal arterial vessels, 18 (30%) had slightly constricted blood vessels, and 10 (16.7%) presented with significant constriction. A weak correlation between age and retinal artery constriction was found (r = 0.345, p = 0.007) (Figure 4).

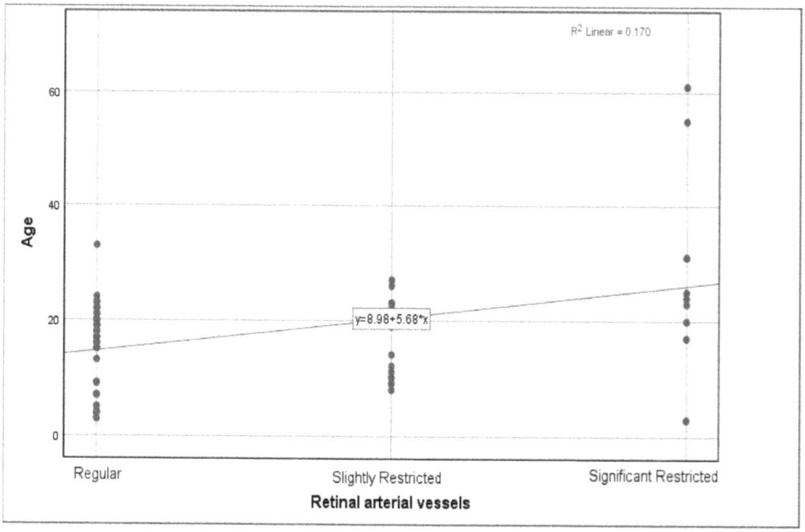

Figure 4. Progressive changes in the retinal arterial vessels as a function of age. A weak relationship between the age and the retinal arterial vessels was found (r = 0.345, p = 0.007). No correlation was found between the mutated gene and the retinal vessels (r = −0.157, p = 0.375).

We noted a weak correlation between the age and the optic nerve head morphology (r = 0.280, p = 0.027).

Out of 61 patients, the majority of 52 (85.2%) had no abnormal findings in the peripheral retina, while 9 (14.8%) presented deposits around the pigment epithelium and other changes in the fundus periphery. No correlation was found between age and the fundus periphery (r = 0.057, p = 0.660) nor between the mutated gene and the fundus periphery (r = −0.135, p = 0.438).

For a sample of 62 patients, 23 (37.1%) did not show changes in the macula, and 20 (32.3%) had a well-delimited wall reflex but not a well-delimited foveal reflex. For 16 (25.8%), the macula was not differentiated, and for the remaining 3 (4.8%), the macula was not differentiated with pigment epithelium alterations. We noted no correlation between age and changes in the macula (r = 0.121, $p = 0.348$) and no correlation between the mutated gene and the macula (r = −0.167, $p = 0.338$).

The majority of the 62 patients ($n = 40$, 64.5%) did not present with any pathology of the optic nerve head, while 7 (11.3%) presented with a slightly pale optic nerve head, 3 (4.8%) presented with a noticeably pale optic nerve head, 7 (11.3%) had pronounced retinal optic atrophy and 5 presented nerve head drusen (8.1%). We noted a weak correlation between age and the optic nerve head (r = 0.280, $p = 0.027$) but noted no correlation between the mutated gene and the optic nerve head (r = −0.249, $p = 0.148$).

3.6. *Quality of Life*

3.6.1. Handicap and Allowances

Achromats are assigned a visual handicap ranging from 0% to 100% according to German regulations for visually impaired citizens. Usually, the amount of allowance received depends on the degree of the handicap. Out of 61 patients, 17 (27.9%) had a 100% degree of visual handicap, 10 (16.4%) had a 90% degree of visual handicap, 14 (23%) had an 80% degree of visual handicap, 9 (14.8%) had a 70% degree of visual handicap, 1 (1.6%) had a 50% degree of visual handicap and 10 (16.4%) had no degree of visual handicap. This confirmed our clinical impression that the degree of visual handicap could be judged very differently legally in spite of a comparable visual handicap. There was no significant relationship between the degree of handicap as a function of age ($n = 61$, r = 0.140, $p = 0.280$). Out of 67 patients, 10 (14.9%) received allowances for the blind, while 57 (85.1%) did not receive any.

3.6.2. Mean Age of Diagnosis, School Career and Academic Achievements

We ranked academic achievements based on the educational benchmarks completed, such as a high school diploma or university degree. We analysed the relationship between the academic achievements of achromats with the age of diagnosis and their attendance at mainstream schools or for those visually handicapped students. The mean age when the diagnosis of achromatopsia was established was 6.49 years (SD ± 1.50), with the oldest patient, age 74, diagnosed at the age of 71 and the youngest patient having been diagnosed shortly after birth. Out of 56 respondents, the majority went to a mainstream school ($n = 40$; 71.4%), while fewer went to schools for the visually impaired ($n = 9$; 16.1%) or a mix of specialised and mainstream schools ($n = 7$; 12.5%).

We received data on the academic qualifications from 28 respondents and noted that 26 had a high school diploma (92.9%), 1 (3.6%) did not have a school qualification, and 1 (3.6%) obtained a middle school diploma. Regarding further education, we received 26 responses, which consisted of one (3.8%) having an apprenticeship, another (3.8%) doing higher intermediate service, and the majority pursuing university studies ($n = 24$; 92.3%). The participants had different job occupations, most working in social professions. No correlation was found between the age of diagnosis established with a school qualification (r = 0.273, $p = 0.168$) or further education (r = −0.335, $p = 0.102$). We did not find any correlation between the type of school and the kind of school qualification achieved (r = −0.124, $p = 0.530$) and between the type of school and further education (r = −0.206, $p = 0.313$).

3.6.3. Use of Optical Aids and Low Vision Aids

For present optical and low vision aids, out of 87 subjects, 27 (31%) had optical aids, 55 (63.2%) used a combination of optical and low vision aids, and 5 (5.7%) had none. Many patients used several types of edge filter glasses following individual customisation and fitting (Tables 3 and 4).

Table 3. Relative proportions of optical aids used $n = 87$ patients.

	Number out of 87 Patients	% out of 87 Patients
Optical aids (for refractive error and glare sensitivity)		
Edge filter glasses with diopter correction	38	43.7%
Edge filter glasses without diopter correction	37	42.5%
Tinted glasses without edge filters, usually sunglasses	22	25.3%
Edge filter contact lenses with diopter correction	13	14.9%
Side Protection	10	11.5%
Contact lenses with diopter values for optical correction	8	9.2%
Edge filter contact lenses without diopter correction	8	9.2%
Overspecs	2	2.3%

Table 4. Relative proportions of low vision aids used $n = 87$ patients.

Low Vision Aids		
Reading glasses	49	56.3%
Simple magnifying glass	48	55.2%
Smartphone	34	39.1%
Tablet	33	37.9%
Monocular	32	36.8%
CCTV (close circuit TV)	27	31%
PC	15	17.2%
Magnifying app	13	14.9%
Zoom-Text	13	14.9%
Screen magnifier	12	13.8%
Electronic magnifying glass	10	11.5%

3.6.4. Primary Impairing Issue of Achromatopsia (Low Vision, Glare Sensitivity, No Colour Vision) of Patients with Achromatopsia

For the question 'What is your primary concern?', out of 43 respondents, the primary wish of 26 (60.5%) achromats was normal visual acuity. In total, 15 (34.9%) achromats wished for no more glare sensitivity, and only 2 (4.7%) wanted intact colour vision.

This was corroborated by the response from 44 respondents concerning the use of a colour system: 30 did not use any, 7 used colorADD, and 7 used a different colour system.

4. Discussion

Our findings suggest that age does not influence visual acuity and that visual acuity usually does not change with age (mean = 1.0 logMAR). These findings are in accordance with most prior studies [20,22,23,25,28,30], which have also concluded that visual acuity becomes stable over time but differs from a few other studies that concluded a decline in visual acuity [21,24]. Refractive error and nystagmus were common traits in our patients, with hyperopia being the most common refractive error. The data we collected indicated no correlation between the degree of spherical refraction and astigmatism against visual acuity.

Regarding ocular morphology, we observed a subtle constriction of the retinal arterial vessels with age, which could be a feature found in achromatopsia [3]. A constriction of the qualitative calibre of the retinal vasculature could, however, also be due to age. The optic nerve also presented morphological changes with age, suggesting retinal atrophy. We found no relationship between fundus periphery changes with age and between macula changes and age. These findings suggest that although progressive retinal changes could be seen, they are likely to develop slowly. This correlates to prior studies [25–28], which observed minor macular changes over time and demonstrated that achromatopsia is a predominantly stationary disorder.

The majority of patients with available genetic data had either a change in the *CNGA3* or the *CNGB3* gene. This finding corresponds to a past study on achromatopsia in which these genes accounted for more than 70% of cases [31]. We did not observe any phenotype–genotype relationship, indicating that the clinical characteristics for genetic mutations of *CNGA3* and *CNGB3* were similar, which is in accordance with past studies [20,21,24,26–28].

With the mean age of diagnosis being established at 6.49 years old, we can say that although the symptoms of achromatopsia were present at birth or early infancy [1], it is often diagnosed later in childhood. Most of the participants were diagnosed by specialised eye departments. Two of our patients even self-diagnosed (internet research) and were later confirmed after consulting an eye doctor who did not identify the disease, indicating that the disease is still not well known among non-specialised ophthalmologists. However, we noted that the diagnosis was found at a later age in older patients. We can, therefore, say that achromatopsia has only recently become better known and is now diagnosed much earlier than in the past. Most of the participants went to regular school, obtained a high school diploma, and attended university, which shows that achromats are not academically hindered. We also noted that attending schools for the visually impaired is not correlated with better or worse academic achievement in this patient population.

Additionally, an earlier age of diagnosis did not influence education, confirming that achromats are not academically handicapped. However, it would be misleading to conclude that achromatopsia is not a handicap in society. There might be a limitation in our study as there could be a greater tendency to answer the questionnaire by academically more proficient people.

The fact that the degree of handicap varied among our patients with a very uniform visual acuity shows that the given degree of handicap is often inappropriate. It should be a goal in the legal classification of (any) handicap to treat and classify patients only following standardised tests and personality-independent examinations. Many participants have indicated that they have problems obtaining aids or allowances for the blind because institutions only consider visual acuity for the degree of handicap in achromats and not colour vision deficits and debilitating photophobia.

According to our results, most patients with achromatopsia need to use optical aids for their symptoms, highlighting the importance of an early and correct diagnosis to manage the symptoms with the appropriate aids. These visual aids have to be fitted in a personalised matter to alleviate the individual symptoms of each patient. Edge filter glasses are the most used, with contact lenses being less used among our patients. This could be explained by the fact that edge filters combined with contact lenses are not accepted by German health insurance regulations and can, therefore, be quite expensive. Additionally, many of the participants indicated in the survey that they would try to avoid edge filter glasses outdoors to avoid social discrimination, showing that achromats try to adapt to society. Hence, contact lenses might be a practical option but costly. Reading glasses and a simple magnifying glass are the most used low-vision aids as they are easier to use than electronic devices. Many patients use these new technologies (smartphones, tablets) as low-vision aids.

Normal visual acuity is the first wish for most patients, followed by a reduction in glare sensitivity and colour vision. These findings are in accordance with Aboshiha et al. [15], who also found out that good visual acuity was the main wish for most achromats, followed by glare sensitivity and colour vision. Furthermore, most participants did not use a colour recognition system. These findings indicate that colour is not the most debilitating symptom in total colour blindness but visual acuity and glare sensitivity are contrary to what one might expect.

In summary, our study suggests that achromatopsia is a relatively stable disease with only subtle morphological changes. Concerning the quality of life, people need to be sensitised, as the handicap arising from achromatopsia is still ambiguous. To our knowledge, this study has the largest sample size for this rare disease with a wide range of ages. Thus, we could make critical statistical comparisons within the group. It was also the first to assess the overall quality of life of patients living with achromatopsia. The limitation was the retrospective aspect with missing information in medical charts, non-respondents to the survey, and subjective answers, which could lead to missing key data and subjective bias.

A personalised approach in achromatopsia is essential when evaluating the need for individual low vision aids and in testing the kind of edge filter. Only this personalised approach can ensure the best development for children and the best working conditions for adults.

In our study, the natural history of achromatopsia is based only on a limited 13-year period in most participants of a younger age. Future longitudinal studies with an extended follow-up period and the inclusion of older patients are needed to strengthen the claim that achromatopsia is a stationary disease. Furthermore, studies with a large sample size are needed to confirm our findings and further analyse individual visual aids and achromats' quality of life.

Author Contributions: Conceptualisation, B.K.-K. and C.C.; Methodology, C.C.; Formal Analysis, B.K.-K.; Investigation, C.C.; Data Curation, C.C.; Writing—original Draft Preparation, C.C.; Writing—review and Editing, B.K.-K. and C.C.; Visualisation, C.C. and B.K.-K.; Supervision, B.K.-K. and B.S. All authors have read and agreed to the published version of the manuscript.

Funding: This research received no external funding.

Institutional Review Board Statement: The study was conducted in accordance with the Declaration of Helsinki and approved by the Ethics Committee of Saarland (protocol code 85/21, 22 March 2021) for studies involving humans.

Informed Consent Statement: Informed consent was obtained from all subjects involved in the study.

Data Availability Statement: The data that support the findings of this study are available from the corresponding author upon reasonable request.

Acknowledgments: The authors would like to thank Haris Sideroudi, who conducted the analysis data and produced the graphs and figures. Furthermore, we would like to thank Hans-Werner Merkelbach (Chairman of Achromatopsie Selbsthilfe e.V. Germany), who distributed the survey to members of the achromatopsia self-help group (www.achromatopsie.de) and to our patients, who contributed to the study.

Conflicts of Interest: The authors declare no conflict of interest.

References

1. Aboshiha, J.; Dubis, A.M.; Carroll, J.; Hardcastle, A.J.; Michaelides, M. The cone dysfunction syndromes. *Br. J. Ophthalmol.* **2015**, *100*, 115–121. [CrossRef]
2. Remmer, M.H.; Rastogi, N.; Ranka, M.P.; Ceisler, E.J. Achromatopsia: A review. *Curr. Opin.* **2015**, *26*, 333–340. [CrossRef]
3. Achromatopsia. GeneReviews®. Available online: https://www.ncbi.nlm.nih.gov/books/NBK1418/ (accessed on 25 May 2021).
4. Tsang, S.H.; Sharma, T. Rod Monochromatism (Achromatopsia). *Adv. Exp. Med. Biol.* **2018**, *1085*, 119–123. [CrossRef] [PubMed]

5. Thomas, M.G.; Papageorgiou, E.; Kuht, H.J.; Gottlob, I. Normal and abnormal foveal development. *Br. J. Ophthalmol.* **2022**, *106*, 593–599. [CrossRef] [PubMed]
6. Solaki, M.; Baumann, B.; Reuter, P.; Andreasson, S.; Audo, I.; Ayuso, C.; Balousha, G.; Benedicenti, F.; Birch, D.; Bitoun, P.; et al. Comprehensive variant spectrum of the CNGA3 gene in patients affected by achromatopsia. *Hum. Mutat.* **2022**, *43*, 832–858. [CrossRef]
7. Weisschuh, N.; Sturm, M.; Baumann, B.; Audo, I.; Ayuso, C.; Bocquet, B.; Branham, K.; Brooks, B.P.; Catalá-Mora, J.; Giorda, R.; et al. Deep-intronic variants in CNGB3 cause achromatopsia by pseudoexon activation. *Hum. Mutat.* **2020**, *41*, 255–264. [CrossRef] [PubMed]
8. Felden, J.; Baumann, B.; Ali, M.; Audo, I.; Ayuso, C.; Bocquet, B.; Casteels, I.; Garcia-Sandoval, B.; Jacobson, S.G.; Jurklies, B.; et al. Mutation spectrum and clinical investigation of achromatopsia patients with mutations in the GNAT2 gene. *Hum. Mutat.* **2019**, *40*, 1145–1155. [CrossRef]
9. Andersen, M.K.G.; Bertelsen, M.; Grønskov, K.; Kohl, S.; Kessel, L. Genetic and Clinical Characterization of Danish Achromatopsia Patients. *Genes* **2023**, *14*, 690. [CrossRef]
10. Weisschuh, N.; Stingl, K.; Audo, I.; Biskup, S.; Bocquet, B.; Branham, K.; Burstedt, M.S.; De Baere, E.; De Vries, M.J.; Golovleva, I.; et al. Mutations in the gene PDE6C encoding the catalytic subunit of the cone photoreceptor phosphodiesterase in patients with achromatopsia. *Hum. Mutat.* **2018**, *39*, 1366–1371. [CrossRef]
11. Kohl, S.; Zobor, D.; Chiang, W.C.; Weisschuh, N.; Staller, J.; Menendez, I.G.; Chang, S.; Beck, S.C.; Garrido, M.G.; Sothilingam, V.; et al. Mutations in the unfolded protein response regulator ATF6 cause the cone dysfunction disorder achromatopsia. *Nat. Genet.* **2015**, *47*, 757–765. [CrossRef]
12. Poloschek, C.; Kohl, S. Achromatopsie. *Ophthalmologe* **2010**, *107*, 571–582. [CrossRef]
13. Michaelides, M.; Hunt, D.; Moore, A. The cone dysfunction syndromes. *Br. J. Ophthalmol.* **2004**, *88*, 291–297. [CrossRef]
14. Schornack, M.; Brown, W.; Siemsen, D. The use of tinted contact lenses in the management of achromatopsia. *Optometry* **2007**, *78*, 17–22. [CrossRef]
15. Aboshiha, J.; Kumaran, N.; Kalitzeos, A.; Hogg, C.; Rubin, G.; Michaelides, M. A Quantitative and Qualitative Exploration of Photoaversion in Achromatopsia. *Investig. Ophthalmol. Vis. Sci.* **2017**, *58*, 3537–3546. [CrossRef]
16. Käsmann-Kellner, B.; Hoffmann, M.B. Achromatopsia—Clinic, diagnostics, genes, brain and quality of life. *Die Ophthalmol.* **2023**, *in press*.
17. Fischer, M.D.; Michalakis, S.; Wilhelm, B.; Zobor, D.; Muehlfriedel, R.; Kohl, S.; Weisschuh, N.; Ochakovski, G.A.; Klein, R.; Schoen, C.; et al. Safety and Vision Outcomes of Subretinal Gene Therapy Targeting Cone Photoreceptors in Achromatopsia. *JAMA Ophthalmol.* **2020**, *138*, 643–651. [CrossRef]
18. Reichel, F.F.; Michalakis, S.; Wilhelm, B.; Zobor, D.; Muehlfriedel, R.; Kohl, S.; Weisschuh, N.; Sothilingam, V.; Kuehlewein, L.; Kahle, N.; et al. Three-year results of phase I retinal gene therapy trial for CNGA3-mutated achromatopsia: Results of a non randomised controlled trial. *Br. J. Ophthalmol.* **2021**, *106*, 1567–1572. [CrossRef]
19. Mendell, J.R.; Al-Zaidy, S.A.; Rodino-Klapac, L.R.; Goodspeed, K.; Gray, S.J.; Kay, C.N.; Boye, S.L.; Boye, S.E.; George, L.A.; Salabarria, S. Current Clinical Applications of In Vivo Gene Therapy with AAVs. *Mol. Ther.* **2021**, *29*, 464–488. [CrossRef]
20. Thiadens, A.A.; Somervuo, V.; van den Born, L.I.; Roosing, S.; van Schooneveld, M.J.; Kuijpers, R.W.; van Moll-Ramirez, N.; Cremers, F.P.; Hoyng, C.B.; Klaver, C.C. Progressive Loss of Cones in Achromatopsia: An Imaging Study Using Spectral-Domain Optical Coherence Tomography. *Investig. Ophthalmol. Vis. Sci.* **2010**, *51*, 5952–5957. [CrossRef]
21. Thiadens, A.A.; Slingerland, N.W.; Roosing, S.; van Schooneveld, M.J.; van Lith-Verhoeven, J.J.; van Moll-Ramirez, N.; van den Born, L.I.; Hoyng, C.B.; Cremers, F.P.; Klaver, C.C. Genetic Etiology and Clinical Consequences of Complete and Incomplete Achromatopsia. *Ophthalmology* **2009**, *116*, 1984–1989. [CrossRef]
22. Thomas, M.G.; Kumar, A.; Kohl, S.; Proudlock, F.A.; Gottlob, I. High-Resolution In Vivo Imaging in Achromatopsia. *Ophthalmology* **2011**, *118*, 882–887. [CrossRef] [PubMed]
23. Thomas, M.G.; McLean, R.J.; Kohl, S.; Sheth, V.; Gottlob, I. Early signs of longitudinal progressive cone photoreceptor degeneration in achromatopsia. *Br. J. Ophthalmol.* **2012**, *96*, 1232–1236. [CrossRef] [PubMed]
24. Brunetti-Pierri, R.; Karali, M.; Melillo, P.; Di Iorio, V.; De Benedictis, A.; Iaccarino, G.; Testa, F.; Banfi, S.; Simonelli, F. Clinical and Molecular Characterization of Achromatopsia Patients: A Longitudinal Study. *Int. J. Mol. Sci.* **2021**, *22*, 1681. [CrossRef] [PubMed]
25. Sundaram, V.; Wilde, C.; Aboshiha, J.; Cowing, J.; Han, C.; Langlo, C.S.; Chana, R.; Davidson, A.E.; Sergouniotis, P.I.; Bainbridge, J.W.; et al. Retinal Structure and Function in Achromatopsia. *Ophthalmology* **2014**, *121*, 234–245. [CrossRef] [PubMed]
26. Genead, M.A.; Fishman, G.A.; Rha, J.; Dubis, A.M.; Bonci, D.M.O.; Dubra, A.; Stone, E.M.; Neitz, M.; Carroll, J. Photoreceptor Structure and Function in Patients with Congenital Achromatopsia. *Investig. Ophthalmol. Vis. Sci.* **2011**, *52*, 7298–7308. [CrossRef]
27. Aboshiha, J.; Dubis, A.M.; Cowing, J.; Fahy, R.T.; Sundaram, V.; Bainbridge, J.W.; Ali, R.R.; Dubra, A.; Nardini, M.; Webster, A.R.; et al. A Prospective Longitudinal Study of Retinal Structure and Function in Achromatopsia. *Investig. Ophthalmol. Vis. Sci.* **2014**, *55*, 5733–5743. [CrossRef]
28. Hirji, N.; Georgiou, M.; Kalitzeos, A.; Bainbridge, J.W.; Kumaran, N.; Aboshiha, J.; Carroll, J.; Michaelides, M. Longitudinal Assessment of Retinal Structure in Achromatopsia Patients with Long-Term Follow-up. *Investig. Ophthalmol. Vis. Sci.* **2018**, *59*, 5735–5744. [CrossRef]

29. Stoianov, M.; de Oliveira, M.S.; dos Santos Ribeiro Silva, M.C.L.; Ferreira, M.H.; de Oliveira Marques, I.; Gualtieri, M. The impacts of abnormal color vision on people's life: An integrative review. *Qual. Life Res.* **2019**, *28*, 855–862. [CrossRef]
30. Georgiou, M.; Singh, N.; Kane, T.; Zaman, S.; Hirji, N.; Aboshiha, J.; Kumaran, N.; Kalitzeos, A.; Carroll, J.; Weleber, R.G.; et al. Long-Term Investigation of Retinal Function in Patients with Achromatopsia. *Investig. Ophthalmol. Vis. Sci.* **2020**, *61*, 38. [CrossRef]
31. Michalakis, S.; Gerhardt, M.; Rudolph, G.; Priglinger, S.; Priglinger, C. Achromatopsia: Genetics and Gene Therapy. *Mol. Diagn. Ther.* **2022**, *26*, 51–59. [CrossRef]

Journal of *Personalized Medicine*

MDPI

Article

Balloon Dacryocystoplasty with Pushed Monocanalicular Intubation as a Primary Management for Primary Acquired Nasolacrimal Duct Obstruction

Chun-Chieh Lai [1,2], Cheng-Ju Yang [3], Chia-Chen Lin [1] and Yi-Chun Chi [4,*]

1 Department of Ophthalmology, National Cheng Kung University Hospital, College of Medicine, National Cheng Kung University, Tainan 704, Taiwan
2 Institute of Clinical Medicine, College of Medicine, National Cheng Kung University, Tainan 704, Taiwan
3 Department of Surgery, Chung Shan Medical University Hospital, Taichung 402, Taiwan
4 Department of Ophthalmology, Kaohsiung Medical University Hospital, Kaohsiung Medical University, Kaohsiung 807, Taiwan
* Correspondence: jenniferchi1308@gmail.com

Abstract: Given the improvement in the instrument and techniques, novel surgical interventions emerged to avoid the osteotomy from the gold standard dacryocystorhinostomy (DCR) for treating primary acquired nasolacrimal duct obstruction (PANDO). This study's aim is to compare the surgical outcomes of antegrade balloon dacryocystoplasty (DCP) with pushed monocanalicular intubation (MCI) to balloon DCP alone in patients with complete PANDO. Adult patients with complete PANDO receiving balloon DCP followed by pushed MCI or balloon DCP alone from December 2014 to May 2019 were retrospectively reviewed. A total of 37 eyes of 29 patients were treated with balloon DCP with pushed MCI for 1 month, whereas 35 eyes of 28 patients were treated with balloon DCP alone. The success rates at 1 month, 3 months, and 6 months after operation were 89.2%, 73.0%, and 70.2%, respectively, in balloon DCP with MCI group, and 62.9%, 62.9%, and 60.0%, respectively, in the balloon DCP alone group. The balloon DCP with pushed MCI group had a better success rate but only reached statistical significance at 1 month postoperatively ($p < 0.01$). Subgroup analysis was performed based on age. The success rate in those under 65 in the combined balloon DCP with MCI group was significantly higher than in balloon DCP alone group (72.7% vs. 9.1%, $p = 0.004$), whereas there was no significant difference between those aged at least 65 in the combined group and the balloon DCP alone group (69.2% vs. 83.3%, $p = 0.2$). Conclusively, there was no significant difference in the success rate between antegrade balloon DCP with and without pushed MCI in general. Nevertheless, the former procedure was associated with significantly higher surgical success rate than the latter in younger patients.

Keywords: balloon dacryocystoplasty; monocanalicular intubation; Masterka tube; nasolacrimal duct obstruction; silicone tube intubation

Citation: Lai, C.-C.; Yang, C.-J.; Lin, C.-C.; Chi, Y.-C. Balloon Dacryocystoplasty with Pushed Monocanalicular Intubation as a Primary Management for Primary Acquired Nasolacrimal Duct Obstruction. *J. Pers. Med.* **2023**, *13*, 564. https://doi.org/10.3390/jpm13030564

Academic Editors: Chieh-Chih Tsai and Yousif Subhi

Received: 31 December 2022
Revised: 12 March 2023
Accepted: 20 March 2023
Published: 21 March 2023

1. Introduction

Acquired nasolacrimal duct obstruction is the blockage of the lacrimal system, resulting in epiphora or excessive tearing, intermittent mucopurulent discharge, and mattering of the involved eye. It occasionally causes lacrimal sac mucocele, fistula, abscess, acute or chronic dacryocystitis, or even orbital cellulitis and cavernous sinus thrombosis [1]. It can be roughly classified as primary acquired nasolacrimal duct obstruction (PANDO) or secondary acquired lacrimal duct obstruction (SALDO) [1]. SALDO is secondary to various defined etiologies of lacrimal obstructions categorized into infectious, inflammatory, traumatic, mechanical, and neoplastic SALDO, and the treatments targeting the cause of the obstruction may efficiently relieve the symptoms, whereas PANDO is defined as nasolacrimal duct obstruction of unknown etiology.

Surgical interventions for PANDO can be simply divided into reconstructive surgery and bypass surgery [2]. Although dacryocystorhinostomy (DCR), which is categorized as bypass surgery, has been the gold standard treatment for PANDO due to its high success rate [3–8], there have been some attempts to avoid osteotomy and perform less invasive procedures to manage nasolacrimal duct obstruction, including balloon dacryocystoplasty (DCP) and silicone stent intubation, which are categorized as reconstructive surgeries. Current studies on balloon DCP, silicone stent intubation, or a combination of the two procedures are mostly focused on partial PANDO, where the reported success rates vary, ranging from 25% to 76% [9–15]. Some studies have discussed balloon DCP for complete PANDO [13,14,16], while limited studies have discussed silicone stent intubation [2,16,17], and even fewer studies reported the result of combination procedure in complete PANDO.

Nowadays, there are two main types of monocanalicular intubation (MCI) available for management of the nasolacrimal duct obstruction. One is a pulled-type MCI and the other is a pushed-type MCI. Compared with the pushed-type, the probe guide should be drawn out from the nostril through the inferior meatus in the pulled-type MCI, which potentially may lead to intra-operative bleeding due to the penetration of the mucosa of inferior meatus, and subsequent patients' discomfort during and after the operation [18]. On the other hand, the probe guide is placed into a silicone tube in the pushed-type MCI, which may require no manipulation of the nostril and therefore decreases the risk of nasal bleeding during the procedure [19]. We therefore employed the pushed-type MCI following balloon DCP in the current study because we believe that it is easier to control and carries a lower risk than pulled-type MCI in conjunction with balloon DCP. In this study, we aimed to compare the surgical outcomes of antegrade balloon DCP with pushed-type MCI to balloon DCP alone in patients with complete PANDO.

2. Materials and Methods

We conducted a retrospective, non-randomized, comparative consecutive study in the Ophthalmology Department of National Cheng Kung University Hospital in Taiwan. Approval from the Institutional Review Board at National Cheng Kung University Hospital was obtained. All procedures in this study with human participants were in accordance with the ethical standards of National Cheng Kung University Hospital, the 1964 Helsinki declaration, its later amendments, and all laws in Taiwan. An informed consent was waived by the Institutional Review Board at National Cheng Kung University Hospital due to the retrospective nature of the study.

From December 2014 to May 2019, patients with complete nasolacrimal duct obstruction (NLDO) based on obstructed intra-sac irrigation and the symptom of epiphora were enrolled in the present study. Post-operative follow up was required to be more than 6 months. Thereafter, we excluded patients aged younger than 18 years old, and those with previous lacrimal procedure, prior ocular adnexal procedures, upper lacrimal drainage obstruction, punctal stenosis, or ocular infection. All patients underwent antegrade balloon DCP (LacriCATH®, QUEST Medical Inc., Allen, TX, USA, 3 mm balloon diameter, 15 mm in length) followed by a pushed monocanalicular silicone stent (Masterka®; FCI SAS, Paris, France) intubation for 1 month or balloon DCP alone as the primary treatment for complete PANDO. Balloon DCP was covered by National Health Insurance in Taiwan whereas pushed-type MCI was not and required an extra fee from patients for the MCI tube. Thus, surgical interventions whether to use the pushed-type MCI or not following balloon DCP in the current study was mainly determined by the patient's decision. Subgroup analysis was performed based on age.

All procedures were performed by the same ophthalmologist, Dr. CCL, under local anesthesia at National Cheng-Kung University Hospital, Taiwan. The procedure of balloon DCP in the two groups was as follows. Initially, we dilated the upper punctum. Then, probing was performed with a Bowman probe. In the next step, we inserted a balloon catheter via the superior punctum until the superior mark of the catheter reached the lacrimal punctum (15 mm distance from the balloon). Next, we inflated the balloon using a manometer at eight atmospheres for 90 s followed by another one at the same pressure for 60 s. After the balloon was deflated, we withdrew the balloon catheter until the second mark of the catheter reached the lacrimal punctum (10 mm from the balloon). A 90-s inflation, deflation, and a 60-s inflation were performed again at eight atmospheres.

In the balloon DCP with MCI group, we then performed the tube-inserting procedures. After the balloon catheter was deflated and removed, we dilated the lower punctum and inserted a sizer through the inferior punctum until its tail reached the nasal floor, meanwhile testing whether the pushed-type antegrade MCI was appropriate for the patient. Next, we measured the proper length of the introducer (30, 35, 40 mm) according to the point on the sizer where the punctum was fitted. After measuring the proper length, we removed the sizer and performed a pushed-type antegrade MCI through the inferior punctum. Then, we withdrew the guide along the axis of the lacrimal sac. Finally, an anchoring plug was inserted into the vertical canaliculus. We placed the pushed-type MCI through the lower punctum under the assumption that more tear flow drains down the lower punctum by gravity. We thought that by placing the pushed-type MCI in the lower rather than upper punctum, the drainage would be improved more efficiently after the removal of the tube. Nevertheless, this assumption has not been confirmed so far.

All operated eyes were treated with topical 0.3% gentamicin and 0.1% betamethasone sodium phosphate four times per day for 2 weeks, followed by topical 4% sulfamethoxazole and 0.1% fluorometholone four times per day for 2 weeks. Postoperative follow up was arranged at approximately 1 week postoperatively. After that, we arranged an appointment at 1 month postoperatively to remove the stents at the outpatient department without anesthesia. Then, we performed intra-sac irrigation to check whether the lacrimal system was patent. Afterwards, we arranged a further outpatient department follow up at 3 months and 6 months postoperatively. Intra-sac irrigation was performed at the two follow-up appointments.

Surgical success was defined as a patent lacrimal draining system based on the intra-sac irrigation results and patient's subjective improvement of symptoms. The surgical outcomes were recorded at 1 month, 3 months, and 6 months postoperatively. Patients who missed their postoperative appointments for intra-sac irrigation and arranged another outpatient department follow-up themselves were included in this study. If the intra-sac irrigation demonstrated patency over 6 months postoperatively, we defined it as surgical success at 3 months and 6 months postoperatively. If the patient was unable to tolerate epiphora and underwent another surgical intervention less than 6 months postoperatively, the outcome was considered a surgical failure at 6 months postoperatively.

A statistical analysis was completed using Mann–Whitney U tests and the Fisher's exact test via software Statistical Product and Service Solutions (SPSS, Armonk, NY, USA), version 20.00. Differences were considered statically significant if the p value was less than 0.05.

3. Results

A total of 72 eyes of 56 patients with complete NLDO treated with balloon DCP with or without pushed type MCI were included in our study. A total of 37 eyes of 29 patients were treated with balloon DCP combined with pushed-type MCI, while 35 eyes of 28 patients were treated with balloon DCP alone. There were no intraoperative complications found in either group. The baseline characteristics of the two groups are provided in Table 1. No significant differences between age, gender, or laterality were found. There were also no between-group differences found in the proportion of patients younger than 65 years old and those at least 65 years old.

At 1 month postoperatively, the balloon DCP with MCI group had a significantly higher success rate than the balloon DCP alone group (89.2% vs. 62.9%, $p = 0.009$). However, no significant differences were found at 3 months (73.0% vs. 62.9%, $p = 0.25$) or 6 months (70.2% vs. 60.0%, $p = 0.25$) postoperatively despite the fact that the success rates of the balloon DCP with MCI group were higher than that of the balloon DCP alone group (Table 2). No postoperative complications were found in patients treated with balloon DCP and MCI; nevertheless, one eye in the balloon DCP alone group developed chronic dacryocystitis.

Table 1. Demographic data in patients in balloon DCP combined with pushed-type MCI group and balloon DCP alone group.

Variable	Balloon DCP with MCI (37 Eyes of 29 Patients)	Balloon DCP Alone (35 Eyes of 27 Patients)	p Value
Gender (male/female)	2/27	6/21	0.14 [a]
Number of eyes			>0.99 [a]
18–64 years old	11	11	
≥65 years old	26	24	
Mean age (years old)	69.3 ± 12.2	66.1 ± 13.2	0.23 [b]
Laterality			0.24 [a]
Right	19	23	
Left	18	12	

[a] Two-tailed Fisher's exact test; [b] Two-tailed Mann—Whitney U test.

Table 2. Postoperative surgical success rate in patients in balloon DCP combined with pushed-type MCI group and balloon DCP alone group.

	Balloon DCP with MCI	Balloon DCP Alone	p Value
1 month	33/37 (89.2%)	22/35 (62.9%)	<0.01 [a]*
3 months	27/37 (73.0%)	22/35 (62.9%)	0.25 [a]
6 months	26/37 (70.2%)	21/35 (60.0%)	0.25 [a]

[a] One-tailed Fisher's exact test * $p < 0.05$.

We conducted a subgroup analysis by dividing the patients into those under 65 years old and those aged at least 65 years old. The demographics and the results of the subgroup analysis are demonstrated in Table 3. No significant differences were found in age, gender, and laterality.

Table 3. Demographic data in patients in balloon DCP combined with pushed-type MCI group and balloon DCP alone group for patients younger than 65 years old and aged at least 65 years old.

	Age <65 Years Old		p Value	Age ≥65 Years Old		p Value
	Balloon DCP with MCI	Balloon DCP		Balloon DCP with MCI	Balloon DCP	
Gender (male/female)	0/10	2/7	0.21 [a]	2/17	4/14	0.40 [a]
Mean age (years)	53.5 ± 8.8	49.8 ± 8.8	0.25 [b]	76.0 ± 5.3	73.6 ± 6.3	0.12 [b]
Laterality			>0.99 [a]			0.27 [a]
Right	6	7		13	16	
Left	5	4		13	8	

[a] Two-tailed Fisher's exact test; [b] Two-tailed Mann–Whitney U test.

The success rates were significantly higher in balloon DCP with MCI group than in balloon DCP alone group in patients aged less than 65 years old at 1 month, 3 months, and 6 months postoperatively (81.8% vs. 9.1%, $p = 0.001$; 72.7% vs. 9.1%, $p = 0.004$; 72.7% vs. 9.1%, $p = 0.004$; Table 4). On the other hand, there were no significant between-group differences in success rate in those treated with balloon DCP with or without MCI in patients at least 65 years old at 1 month, 3 months, and 6 months postoperatively (92.3% vs. 87.5%, $p = 0.46$; 73.1% vs. 87.5%, $p = 0.18$; 69.2% vs. 83.3%, $p = 0.20$; Table 4).

Table 4. Postoperative surgical success rate in patients in balloon DCP combined with pushed-type MCI group and balloon DCP alone group for patients younger than 65 years old and aged at least 65 years old.

	Age <65 Years Old		p Value	Age ≥65 Years Old		p Value
	Balloon DCP with MCI	Balloon DCP		Balloon DCP with MCI	Balloon DCP	
1 month	9/11 (81.8%)	1/11 (9.1%)	<0.01 [a*]	24/26 (92.3%)	21/24 (87.5%)	0.46 [a]
3 months	8/11 (72.7%)	1/11 (9.1%)	<0.01 [a*]	19/26 (73.1%)	21/24 (87.5%)	0.18 [a]
6 months	8/11 (72.7%)	1/11 (9.1%)	<0.01 [a*]	18/26 (69.2%)	20/24 (83.3%)	0.20 [a]

[a] One-tailed Fisher's exact test. * $p < 0.05$.

4. Discussion

The nasolacrimal duct obstruction can be categorized into congenital or acquired according to the onset of age, and the management of the two are fundamentally different. The etiology of congenital nasolacrimal duct obstruction (CNLDO) is mostly the failure of nasolacrimal duct canalization, which is often in regard to mechanical obstruction at the valve of Hasner [20]. CNLDO affects approximately 20% of newborns worldwide [20]. Among them, around 96% of infants of CNLDO have spontaneous resolution of the symptoms in their first year of life [21]. Another study revealed a 60% spontaneous resolution after the children were 2 years old and a decreasing rate of spontaneous remission after they turned 18 months old [22]. Therefore, the current consensus of treating CNLDO is observation of the symptom of epiphora in the 1 year of age before further surgical intervention for CNLDO. According to our experience, antegrade balloon DCP followed by a short-term intubation with the pushed-type MCI is a high-potential surgical approach with a high success rate (96.77%) as a primary surgical treatment for CNLDO [19].

Acquired nasolacrimal duct obstruction is the blockage of the lacrimal system, resulting in epiphora, intermittent mucopurulent discharge, and mattering of the involved eye. It can be roughly classified as PANDO and SALDO [1]. SALDO is secondary to defined etiologies of obstructions, and is categorized into infectious, inflammatory, traumatic, mechanical, and neoplastic SALDO. The main treatment is to define and target the cause of the obstruction.

PANDO is defined as nasolacrimal duct obstruction of unknown etiology. Different from CNLDO and SALDO, the exact cause of PANDO is under investigation and seems to be multifactorial. PANDO may be a result of chronic inflammation, subsequent fibrosis, and further obstruction of the nasolacrimal duct. Some pathological studies have also shown fibrous formation secondary to chronic inflammation in the obstructed lacrimal drainage system [23]. The chronic inflammation and fibrosis may be the reason that the balloon DCP combined with pushed-type MCI in patients with PANDO had a lower success rate than that in patients with CNLDO. There is other reported pathophysiology of PANDO. Ectopic nasal epithelial cells in the nasolacrimal duct may also play a role in nasolacrimal duct obstruction [24] and the lacrimal duct-associated lymphoid tissue derangement may alter the immune response and could further lead to PANDO [25]. In the current study, we found PANDO mostly affected patients around their sixth and seventh decades, characterized with female predominance, which was compatible with previous studies [7,8]. It might result from the relatively smaller diameter of nasolacrimal duct in female patients [26,27], hormonal changes particularly in postmenopausal women [28], and derangements of lacrimal drainage-associated lymphoid tissue in chronic dacryocystitis [25].

Surgical interventions for PANDO can be divided into reconstructive surgery and bypass surgery [2]. DCR, a bypass surgery, is currently considered as the standard treatment for PANDO due to its high success rate, which was reported with approximately 90% [3,4,6,8]. When it comes to complete PANDO, DCR has also been suggested as the gold standard. A meta-analysis reported a success rate varied from 64 to 100% in external DCR and 84 to 94% in endoscopic DCR [29]. Over the past few decades, endoscopic DCR has grown in popularity. In comparison with external DCR, endoscopic DCR has the advantages of better cosmetic outcome and no facial scarring, a shorter wound recovery time and hospital stay, less blood loss during surgery, and better visualization of endonasal anatomy. Endoscopic DCR enables the operator to simultaneously correct nasal abnormalities like a deviated nasal septum and hypertrophic turbinate, and it is applicable for acute dacryocystitis [30–35]. Disadvantages of endoscopic DCR include more expensive equipment and steep learning curve that a thorough understanding of endonasal anatomy is required [30,32,33]. Our experience in endoscopic DCR with silicone stenting for complete PANDO showed a success rate of 82% in endoscopic DCR with pulled-type MCI and 87% in endoscopic DCR with pushed-type bicanalicular intubation (BCI) [8]. Furthermore, the subgroup analysis of our previous study showed no significant difference of the success rates in patients with or without previous dacryocystitis [8]. Although DCR possesses a better surgical outcome, it resolves the PANDO at the cost of invasiveness, such as osteotomy regardless of endonasal or external approach. In order to decrease the level of invasiveness and bleeding while operating; transcanalicular laser-assisted DCR was developed as an alternative to the conventional DCR by Michalos et al. [36]. However, Ayintap et al. revealed a lower success rate (62%) in transcanalicular laser-assisted DCR in complete PANDO patients younger than 65 years old at 2-year follow up [37].

To avoid osteotomy and decrease the invasiveness, reconstructive surgeries were attempted to manage PANDO. Kuchar and Steinkogler used antegrade balloon DCP combined with silicone stent intubation for 3 to 6 months for complete NLDO, and the success rate was 90% at 3 months and 70% at 6 months postoperatively [16]. In the present study, antegrade balloon DCP was combined with pushed-type MCI intubation for only 1 month, which is the shortest reported intubation duration to date [7,9–12,15–17,38]. We believed that a shorter intubation time would contribute to a lower rate of early stent loss and reduce possible complications related to stents, such as keratitis, canaliculitis, granulation formation, or intracanalicular stent migration. In our study, we placed the pushed-type MCI through the lower punctum into the nasolacrimal duct under the assumption that more tear flow drains down the lower punctum by gravity and, thus, drains more efficiently from the lower punctum. This was different from those described in previous studies by Fayet et al. [39], in which the authors favored inserting the intubation through the upper punctum to prevent the punctal plug of MCI from touching the cornea and causing keratitis. Nonetheless, there was no corneal irritation or keratitis reported after surgery in our study, indicating that the punctal plug of pushed-type MCI anchored on the lower punctum did not raise the risk of corneal damage by stents.

A randomized trial conducted by Andalib et al. compared bicanalicular intubation (BCI) and MCI for partial PANDO, revealing no significant differences in the surgical outcome between the two [15]. However, to date, there have been no comparative studies for complete PANDO. Mimura et al. reported a clinical success rate of 91.7% at 6 months postoperatively using BCI for complete PANDO without canaliculi involvement, while Inatani et al. reported a success rate of 68% [2,17]. However, surgical outcomes for MCI for complete PANDO have not yet been studied.

We favored MCI rather than BCI in the present study. Despite the use of prophylactic antibiotics, the postoperative infection rate with BCI for complete PANDO has been reported in previous studies to be as high as 9.5% [15,17]. Furthermore, a high punctal laceration rate (13.6%) was also reported by Kashkouli et al. [38]. Each case in our study presented with PANDO without canalicular involvements; therefore, bicanalicular stents may be unnecessary. Wladis et al. reported a risk of damage to the uninvolved canaliculus with the BCI [40]. As a result, pushed-type MCI seemed to be a better option in our study due to adequate intubation of the nasolacrimal duct, easy and speedy insertion, and less discomfort for the patients. However, unlike MCI, bicanalicular stents can be fixed without a punctal plug. Therefore, if the surgeons considered the opening of the punctum to be inadequate for the punctal plug to be properly anchored in MCI, the BCI may be the preferred choice.

There are two currently available types of MCI: a pulled-type and a pushed-type MCI. The pulled-type MCI required some interventions at the inferior turbinate while the pushed-type one did not [41]. We favored pushed-type MCI stents because repeated manipulations at the inferior meatus were not required during intubation when comparing with pulled-type MCI or pulled-type BCI. This may reduce operation time, lessen intra-operative nasal bleeding, and minimize pain for patients.

Our success rate at 3 months postoperatively seemed to be lower when compared with a previous study in which antegrade balloon DCP with silicone stent intubation was combined (73.0% vs. 90%, respectively), but the result was similar at 6 months postoperatively (70.2% vs. 70%) [16]. Since a longer postoperative follow-up period has been suggested, the differences between previous studies and the current study at 3 months and 6 months postoperatively were not significant [16]. Furthermore, the current study revealed that the additional pushed-type MCI for antegrade balloon DCP did not significantly improve the surgical outcomes in general since the success rates were not significantly higher at 3 months and 6 months postoperatively despite the significantly higher success rate at 1 month postoperatively.

To our knowledge, our study is the first to compare the differences in success rates between patients younger than 65 years old and those older than 65 for balloon DCP and silicone stent intubation. Interestingly, we found the success rates when combining balloon DCP with pushed-type MCI were always significantly higher than when using balloon DCP alone in patients younger than 65 years old regardless of the postoperative timing. On the other hand, there were no significant differences in the success rates between patients receiving balloon DCP with or without pushed-type MCI if the patients were aged over 65 years old. The silicone stent maintained the patency of the nasolacrimal duct [42], while balloon DCP only creates a perioperatively patent nasolacrimal duct (NLD). The stent intubation might weigh more in younger patients who possess stronger inflammatory reaction and more wound adhesion post-operatively than the older patients. By combining balloon DCP with pushed-type MCI, the patency of the NLD was maintained by avoiding excessive scarring, and consequently resulted in a better surgical outcome.

Due to the invasiveness of the bypass surgery, reconstructive surgery such as balloon DCP or stent intubation also plays a significant role in the management of PANDO. Although our overall long-term success rates with balloon DCP combined with pushed-type MCI (70.2%) did not achieve that using conventional DCR (around 90%), the procedure still yielded several advantages. First, no general anesthesia was required for the surgical method in the current study. Second, we minimized the degree of invasiveness by preserving the original nasolacrimal system instead of performing an osteotomy and largely decreased the operation time. Third, the present method did not require further manipulation of the inferior nasal meatus; thus, we decreased the possibility of significant bleeding. Last but not least, no obvious immediate postoperative complications were found after antegrade balloon DCP with the pushed-type MCI in the current study. Many patients, especially elderly individuals, may select balloon DCP with or without intubations instead of endoscopic DCR as the primary therapy for their nasolacrimal duct blockages due to the benefits indicated above.

In an attempt to improve the surgical success of treating NLDO, antimetabolites such as mitomycin-C have been used in adjunct with probing [43], external DCR [44,45], as well as endoscopic DCR [45], and have resulted in a better success rate with the application of mitomycin-C. However, there has been no consensus regarding the dosage of mitomycin-C to date. Tsai et al. reported that 89% of eyes treated with lacrimal probing with adjunctive low-dose mitomycin-C (0.2 mg/mL) irrigation achieved a patent nasolacrimal duct 9 months after lacrimal probing for epiphora in adults, and no complications were noted during the follow-up period [43]. Various concentrations and durations of mitomycin-C application in DCR were also reported and reviewed [45]. Because of the effect of inhibiting the synthesis of collagen by fibroblasts and subsequently reducing the fibrosis, applying adjunctive low-dose mitomycin-C may be a feasible and safe way to improve the surgical outcome of antegrade balloon DCP with or without stenting in our current study, but further study is needed for a more solid conclusion.

The current study was limited by its retrospective and non-randomized design. Additionally, we did not use any radiographic assistance to confirm the diagnosis of complete PANDO, for example, transcanalicular endoscopy or dacryocystography [13,16]. In addition, our case number was relatively small, which may limit the statistical power. Current results implied efficiency and few complications with antegrade balloon DCP followed by pushed-type MCI for patients with complete PANDO at 6 months postoperatively, and further large-scaled investigations with longer follow-up duration are needed for a more definite long-term result.

In general, the combination of balloon DCP and pushed-type MCI did not lead to a significantly higher success rate as compared with balloon DCP alone for complete PANDO. However, in patients under 65 years old with complete PANDO, the combination of balloon DCP and pushed-type MCI was associated with significantly higher surgical success rates as compared with balloon DCP alone. In addition, the former procedure had several advantages over DCR, including its non-invasiveness and the avoidance of general anesthesia as well as hospitalization for postoperative care. Balloon DCP followed by pushed-type MCI could be of value when it comes to treating complete PANDO in younger patients and it could also serve as an alternative for the elderly who are poor candidates for general anesthesia.

Author Contributions: Conceptualization, C.-C.L. (Chun-Chieh Lai) and C.-C.L. (Chia-Chen Lin); methodology, C.-C.L. (Chun-Chieh Lai) and Y.-C.C.; data acquisition and analysis, C.-J.Y. and C.-C.L. (Chia-Chen Lin); writing—original draft preparation, C.-J.Y.; writing—review and editing, C.-C.L. (Chia-Chen Lin), C.-C.L. (Chun-Chieh Lai) and Y.-C.C. All authors have read and agreed to the published version of the manuscript.

Funding: This research received no external funding.

Institutional Review Board Statement: The study was conducted in accordance with the Declaration of Helsinki and approved by the Institutional Ethics Committee of Institutional Review Board at National Cheng Kung University Hospital, Taiwan. (A-ER-109-139, dated 25 May 2020).

Informed Consent Statement: Patient consent was waived by the Institutional Review Board at National Cheng Kung University Hospital due to the retrospective nature of study.

Data Availability Statement: The datasets in the current study will be available from the corresponding author on reasonable request.

Conflicts of Interest: The authors declare no conflict of interest.

References

1. Kamal, S.; Ali, M.J. Primary acquired nasolacrimal duct obstruction (PANDO) and secondary acquired lacrimal duct obstructions (SALDO). In *Principles and Practice of Lacrimal Surgery*; Springer: Singapore, 2018; pp. 163–171.
2. Mimura, M.; Ueki, M.; Oku, H.; Sato, B.; Ikeda, T. Indications for and effects of Nunchaku-style silicone tube intubation for primary acquired lacrimal drainage obstruction. *Jpn. J. Ophthalmol.* **2015**, *59*, 266–272. [CrossRef] [PubMed]
3. Su, P.-Y. Comparison of endoscopic and external dacryocystorhinostomy for treatment of primary acquired nasolacrimal duct obstruction. *Taiwan J. Ophthalmol.* **2018**, *8*, 19. [CrossRef]
4. Duwal, S.; Saiju, R. Outcomes of external dacryocystorhinostomy and endoscopic endonasal dacryocystorhinostomy in the management of nasolacrimal duct obstruction. *Nepal. J. Ophthalmol.* **2015**, *7*, 39–46. [CrossRef] [PubMed]
5. Jawaheer, L.; MacEwen, C.J.; Anijeet, D. Endonasal versus external dacryocystorhinostomy for nasolacrimal duct obstruction. *Cochrane Database Syst. Rev.* **2017**, *2*, CD007097. [CrossRef] [PubMed]
6. Tarbet, K.J.; Custer, P.L. External dacryocystorhinostomy: Surgical success, patient satisfaction, and economic cost. *Ophthalmology* **1995**, *102*, 1065–1070. [CrossRef]
7. Delaney, Y.; Khooshabeh, R. External dacryocystorhinostomy for the treatment of acquired partial nasolacrimal obstruction in adults. *Br. J. Ophthalmol.* **2002**, *86*, 533–535. [CrossRef]
8. Chi, Y.-C.; Lai, C.-C. Endoscopic dacryocystorhinostomy with short-term, pushed-type bicanalicular intubation vs. pulled-type monocanalicular intubation for primary acquired nasolacrimal duct obstruction. *Front. Med.* **2022**, *9*, 946083. [CrossRef] [PubMed]

9. Kashkouli, M.; Beigi, B.; Tarassoly, K.; Kempster, R. Endoscopically assisted balloon dacryocystoplasty and silicone intubation versus silicone intubation alone in adults with incomplete nasolacrimal duct obstruction. *Eur. J. Ophthalmol.* **2006**, *16*, 514–519. [CrossRef] [PubMed]
10. Bleyen, I.; Van den Bosch, W.A.; Bockholts, D.; Mulder, P.; Paridaens, D. Silicone intubation with or without balloon dacryocystoplasty in acquired partial nasolacrimal duct obstruction. *Am. J. Ophthalmol.* **2007**, *144*, 776–780.e771. [CrossRef]
11. Couch, S.M.; White, W.L. Endoscopically assisted balloon dacryoplasty treatment of incomplete nasolacrimal duct obstruction. *Ophthalmology* **2004**, *111*, 585–589. [CrossRef]
12. Ali, M.J.; Naik, M.N. Efficacy of endoscopic guided anterograde 3 mm balloon dacryoplasty with silicone intubation in treatment of acquired partial nasolacrimal duct obstruction in adults. *Saudi J. Ophthalmol.* **2014**, *28*, 40–43. [CrossRef] [PubMed]
13. Konuk, O.; Ilgit, E.; Erdinc, A.; Onal, B.; Unal, M. Long-term results of balloon dacryocystoplasty: Success rates according to the site and severity of the obstruction. *Eye* **2008**, *22*, 1483–1487. [CrossRef]
14. Lee, J.M.; Song, H.Y.; Han, Y.M.; Chung, G.H.; Sohn, M.H.; Kim, C.S.; Choi, K.C. Balloon dacryocystoplasty: Results in the treatment of complete and partial obstructions of the nasolacrimal system. *Radiology* **1994**, *192*, 503–508. [CrossRef]
15. Andalib, D.; Nabie, R.; Abbasi, L. Silicone intubation for nasolacrimal duct stenosis in adults: Monocanalicular or bicanalicular intubation. *J. Craniofac. Surg.* **2014**, *25*, 1009–1011. [CrossRef]
16. Kuchar, A.; Steinkogler, F.J. Antegrade balloon dilatation of nasolacrimal duct obstruction in adults. *Br. J. Ophthalmol.* **2001**, *85*, 200–204. [CrossRef] [PubMed]
17. Inatani, M.; Yamauchi, T.; Fukuchi, M.; Denno, S.; Miki, M. Direct silicone intubation using Nunchaku-style tube (NST-DSI) to treat lacrimal passage obstruction. *Acta Ophthalmol. Scand.* **2000**, *78*, 689–693. [CrossRef] [PubMed]
18. Elewah, E.-S.M.; Ahmed, A.-E.H.; Ali, A.-M.E.; Ebrahim, A.M. A Comparison between Pulled and Pushed Monocanalicular Silicone Intubation in Management of Congenital Nasolacrimal Duct Obstruction. *Egypt. J. Hosp. Med.* **2018**, *72*, 5215–5220. [CrossRef]
19. Lai, C.C.; Yang, C.J.; Lin, C.C.; Chi, Y.C. Surgical Outcomes of Balloon Dacryocystoplasty Combined With Pushed-Type Monocanalicular Intubation as the Primary Management for Congenital Nasolacrimal Duct Obstruction. *J. Pediatr. Ophthalmol. Strabismus* **2021**, *58*, 365–369. [CrossRef]
20. Sathiamoorthi, S.; Frank, R.D.; Mohney, B.G. Incidence and clinical characteristics of congenital nasolacrimal duct obstruction. *Br. J. Ophthalmol.* **2019**, *103*, 527–529. [CrossRef]
21. MacEwen, C.J.; Young, J.D. Epiphora during the first year of life. *Eye* **1991**, *5 Pt 5*, 596–600. [CrossRef]
22. Young, J.D.; MacEwen, C.J.; Ogston, S.A. Congenital nasolacrimal duct obstruction in the second year of life: A multicentre trial of management. *Eye* **1996**, *10 Pt 4*, 485–491. [CrossRef] [PubMed]
23. Mauriello, J.A., Jr.; Palydowycz, S.; DeLuca, J. Clinicopathologic study of lacrimal sac and nasal mucosa in 44 patients with complete acquired nasolacrimal duct obstruction. *Ophthalmic Plast. Reconstr. Surg.* **1992**, *8*, 13–21. [CrossRef] [PubMed]
24. Paulsen, F.P.; Thale, A.B.; Hallmann, U.J.; Schaudig, U.; Tillmann, B.N. The cavernous body of the human efferent tear ducts: Function in tear outflow mechanism. *Investig. Ophthalmol. Vis. Sci.* **2000**, *41*, 965–970. [CrossRef] [PubMed]
25. Ali, M.J.; Mulay, K.; Pujari, A.; Naik, M.N. Derangements of lacrimal drainage-associated lymphoid tissue (LDALT) in human chronic dacryocystitis. *Ocul. Immunol. Inflamm.* **2013**, *21*, 417–423. [CrossRef] [PubMed]
26. Janssen, A.G.; Mansour, K.; Bos, J.J.; Castelijns, J.A. Diameter of the bony lacrimal canal: Normal values and values related to nasolacrimal duct obstruction: Assessment with CT. *Am. J. Neuroradiol.* **2001**, *22*, 845–850.
27. Groessl, S.A.; Sires, B.S.; Lemke, B.N. An anatomical basis for primary acquired nasolacrimal duct obstruction. *Arch. Ophthalmol.* **1997**, *115*, 71–74. [CrossRef]
28. Ali, M.J.; Schicht, M.; Paulsen, F. Qualitative Hormonal Profiling of the Lacrimal Drainage System: Potential Insights into the Etiopathogenesis of Primary Acquired Nasolacrimal Duct Obstruction. *Ophthalmic Plast. Reconstr. Surg.* **2017**, *33*, 381–388. [CrossRef]
29. Leong, S.C.; MacEwen, C.J.; White, P.S. A systematic review of outcomes after dacryocystorhinostomy in adults. *Am. J. Rhinol. Allergy* **2010**, *24*, 81–90. [CrossRef]
30. Karim, R.; Ghabrial, R.; Lynch, T.; Tang, B. A comparison of external and endoscopic endonasal dacryocystorhinostomy for acquired nasolacrimal duct obstruction. *Clin. Ophthalmol.* **2011**, *5*, 979–989. [CrossRef]
31. Al-Qahtani, A.S. Primary Endoscopic Dacryocystorhinostomy with or without Silicone Tubing: A Prospective Randomized Study. *Am. J. Rhinol. Allergy* **2012**, *26*, 332–334. [CrossRef] [PubMed]
32. Yakopson, V.S.; Flanagan, J.C.; Ahn, D.; Luo, B.P. Dacryocystorhinostomy: History, evolution and future directions. *Saudi J. Ophthalmol.* **2011**, *25*, 37–49. [CrossRef] [PubMed]
33. Amadi, A.J. Endoscopic DCR vs External DCR: What's Best in the Acute Setting? *J. Ophthalmic Vis. Res.* **2017**, *12*, 251–253. [CrossRef]
34. Chong, K.K.-l.; Abdulla, H.A.A.; Ali, M.J. An update on endoscopic mechanical and powered dacryocystorhinostomy in acute dacryocystitis and lacrimal abscess. *Ann. Anat.-Anat. Anz.* **2020**, *227*, 151408. [CrossRef]
35. Kamal, S.; Ali, M.J.; Pujari, A.; Naik, M.N. Primary Powered Endoscopic Dacryocystorhinostomy in the Setting of Acute Dacryocystitis and Lacrimal Abscess. *Ophthalmic Plast. Reconstr. Surg.* **2015**, *31*, 293–295. [CrossRef]
36. Michalos, P.; Pearlman, S.; Avila, E.; Newton, J. Hemispherical tip contact Nd: YAG translacrimal-nasal dacryocystorhinostomy. *Ocul. Surg. News* **1995**, *13*, 40.

37. Ayintap, E.; Buttanri, I.B.; Sadıgov, F.; Serin, D.; Ozsutcu, M.; Umurhan Akkan, J.C.; Tuncer, K. Analysis of age as a possible prognostic factor for transcanalicular multidiode laser dacryocystorhinostomy. *J. Ophthalmol.* **2014**, *2014*, 913047. [CrossRef] [PubMed]
38. Kashkouli, M.B.; Kempster, R.C.; Galloway, G.D.; Beigi, B. Monocanalicular versus bicanalicular silicone intubation for nasolacrimal duct stenosis in adults. *Ophthalmic Plast. Reconstr. Surg.* **2005**, *21*, 142–147. [CrossRef] [PubMed]
39. Fayet, B.; Katowitz, W.R.; Racy, E.; Ruban, J.-M.; Katowitz, J.A. Pushed monocanalicular intubation: An alternative stenting system for the management of congenital nasolacrimal duct obstructions. *J. Am. Assoc. Pediatr. Ophthalmol. Strabismus* **2012**, *16*, 468–472. [CrossRef]
40. Wladis, E.J.; Aakalu, V.K.; Tao, J.P.; Sobel, R.K.; Freitag, S.K.; Foster, J.A.; Mawn, L.A. Monocanalicular Stents in Eyelid Lacerations: A Report by the American Academy of Ophthalmology. *Ophthalmology* **2019**, *126*, 1324–1329. [CrossRef]
41. Fayet, B.; Racy, E.; Ruban, J.-M.; Katowitz, J. Pushed monocanalicular intubation. Pitfalls, deleterious side effects, and complications. *J. Français d'Ophtalmologie* **2011**, *34*, 597–607. [CrossRef]
42. Aritürk, N.; Oüge, I.h.; Öge, F.; Erkan, D.; Havuz, E. Silicone intubation for obstruction of the nasolacrimal duct in adults. *Acta Ophthalmol. Scand.* **1999**, *77*, 481–482. [CrossRef] [PubMed]
43. Tsai, C.-C.; Kau, H.-C.; Kao, S.-C.; Hsu, W.-M.; Liu, J.-H. Efficacy of probing the nasolacrimal duct with adjunctive Mitomycin-C for epiphora in adults. *Ophthalmology* **2002**, *109*, 172–174. [CrossRef]
44. You, Y.A.; Fang, C.T. Intraoperative mitomycin C in dacryocystorhinostomy. *Ophthalmic Plast. Reconstr. Surg.* **2001**, *17*, 115–119. [CrossRef] [PubMed]
45. Nair, A.G.; Ali, M.J. Mitomycin-C in dacryocystorhinostomy: From experimentation to implementation and the road ahead: A review. *Indian J. Ophthalmol.* **2015**, *63*, 335–339. [CrossRef]

Journal of
*Personalized
Medicine*

MDPI

Article

Circadian Fluctuation Changes in Intraocular Pressure Measured Using a Contact Lens Sensor in Patients with Glaucoma after the Adjunctive Administration of Ripasudil: A Prospective Study

Shih-Kung Huang [1,2,*], Mai Ishii [1], Yuki Mizuki [1], Tatukata Kawagoe [1], Masaki Takeuchi [1], Eiichi Nomura [1] and Nobuhisa Mizuki [1]

1 Department of Ophthalmology and Visual Science, Yokohama City University Graduate School of Medicine, Yokohama 236-0004, Kanagawa, Japan
2 Department of Ophthalmology, Yokohama Hodogaya Central Hospital, Yokohama 240-8585, Kanagawa, Japan
* Correspondence: t206029f@yokohama-cu.ac.jp; Tel.: +81-45-787-2683

Abstract: Nocturnal and circadian intraocular pressure (IOP) fluctuations are important issues in glaucoma treatment. Ripasudil 0.4% eye drops, a new glaucoma medication, lowers IOP by increasing aqueous humor outflow through the trabecular meshwork. We aimed to compare differences between circadian IOP fluctuations measured using a contact lens sensor (CLS) before and after administering 0.4% ripasudil eye drops adjunctively to patients with primary open-angle glaucoma (POAG) and normal tension glaucoma (NTG). Patients with POAG ($n = 1$) and NTG ($n = 5$) underwent 24 h IOP monitoring with a CLS before and after administering ripasudil eye drops every 12 h (8 a.m., 8 p.m.) for 2 weeks without discontinuing currently used glaucoma medications. No vision-threatening adverse event occurred. The reduction in IOP fluctuation and the reduction in the SD of IOP in 24 h, awake time and sleep time did not reach statistical significance. The baseline office-hour IOP, which was measured using Goldmann applanation tonometry (GAT), ranged in the low teens, and the reduction in office-hour IOP also did not show a significant difference. Further study is necessary to evaluate whether the low baseline IOP with less IOP reduction relates to attenuated IOP fluctuation reduction.

Keywords: circadian IOP fluctuation; contact lens sensor; glaucoma; intraocular pressure; ripasudil

Citation: Huang, S.-K.; Ishii, M.; Mizuki, Y.; Kawagoe, T.; Takeuchi, M.; Nomura, E.; Mizuki, N. Circadian Fluctuation Changes in Intraocular Pressure Measured Using a Contact Lens Sensor in Patients with Glaucoma after the Adjunctive Administration of Ripasudil: A Prospective Study. *J. Pers. Med.* **2023**, *13*, 800. https://doi.org/10.3390/jpm13050800

Academic Editors: Ana Isabel Ramírez Sebastián and Chieh-Chih Tsai

Received: 24 February 2023
Revised: 23 April 2023
Accepted: 5 May 2023
Published: 6 May 2023

1. Introduction

Glaucoma refers to a group of ocular diseases that can result in optic nerve fiber layer loss leading to constriction of the visual field. Glaucoma is a major cause of irreversible blindness globally. Lowering the intraocular pressure (IOP) remains the most important treatment for glaucoma [1]. Despite achieving good control of IOP, deterioration of the visual field may still be observed in routine clinical practice. Determining whether IOP other than that measured during office hours is well controlled remains a challenge [2]. Measuring IOP several times a day using Goldmann applanation tonometry (GAT) or non-contact pneumotonometry may solve this problem [3,4]. However, GAT and non-contact pneumotonometry are measured in the sitting position, and body position may affect changes in IOP. Moreover, the limited frequency of measurements poses a challenge to the detection of any change in IOP [5]. A contact lens sensor (CLS) can continuously measure the IOP for the 24 h in a day and in the supine position while sleeping; therefore, it could solve this problem [6].

Ripasudil (Glanatec® ophthalmic solution, 0.4%; Kowa Company, Ltd., Nagoya, Japan) is a Rho-associated coiled-coil-containing protein kinase inhibitor drug, and in Japan, it was approved for glaucoma treatment in 2014. Ripasudil lowers IOP by inducing a morphological change in the trabecular meshwork (TM) and then increasing the conventional aqueous outflow through the TM and Schlemm's canal [7]. The IOP-lowering effect of ripasudil has been demonstrated in previous studies [8]. Tanihara et al. reported that the IOP-lowering effect of ripasudil persisted for 7 h after the first (9 a.m.) and second (9 p.m.) administrations [9]. Moreover, ripasudil was more effective in lowering IOP during the day and at night than the placebo. However, the IOP was measured in a sitting position using GAT, and sleep was disrupted during the measurement. The nocturnal IOP-lowering effect of ripasudil in the supine position can be measured using a CLS without disrupting sleep; therefore, we can acquire information regarding the IOP fluctuation change after ripasudil eye drop administration under natural physiological conditions.

To the best of our knowledge, there has been no research to date, and this is the first study to compare the 24 h IOP-fluctuation changes before and after a 2-week adjunctive administration of ripasudil eye drops using a CLS in patients with glaucoma.

2. Materials and Methods

2.1. Patients

This was a prospective open-label, single-arm study. Between January 2022 and April 2023, one patient with primary open-angle glaucoma (POAG) and five patients with normal tension glaucoma (NTG) from the Yokohama Hodogaya Central Hospital, Japan, a co-sponsor institute of this study, were enrolled in this study. The inclusion criteria required the patients to be aged 20 years or older and diagnosed with POAG or NTG. The diagnostic requirements for a glaucomatous visual field included the following: (1) a glaucoma hemifield test result outside the normal limits, compatible with retinal nerve fiber layer (RNFL) loss; (2) three or more abnormal points with a probability of being normal of $p < 5\%$ and at least one point with a pattern deviation of $p < 1\%$; or (3) a pattern standard deviation (SD) of $p < 5\%$. Patients were excluded if (1) their condition was diagnosed as secondary open-angle glaucoma or angle-closure glaucoma, (2) they had any history of silicone allergy or adverse reactions to contact lens use before, or (3) they received systemic steroid treatment or topical steroid eye drops treatment for systemic disease or uveitis. The patients underwent a complete ophthalmic examination during the screening visit, including slit-lamp biomicroscopy, gonioscopy, GAT, refraction tests, central corneal thickness (CCT) measurement, dilated fundus examination, standard automated perimetry (Humphrey 30-2 SITA-Standard; Carl Zeiss Meditec, Inc., Dublin, CA, USA), and RNFL thickness examination (SS-OCT; DRI OCT Triton-1, Topcon, Tokyo, Japan). The eye with the worst mean deviation in the visual field test was recruited. The baseline office-hour IOP was measured using GAT, and the 24 h IOP was measured using a CLS without hospital admission. The CLS was administered by an ophthalmologist in the ophthalmology outpatient department. The patients were instructed to record the time of sleep and the time of awakening. The office-hour GAT IOP was measured before applying a CLS, and then CLS 24 h IOP measurements were repeated at around the same time in the day as the first CLS measurement after the 2-week administration of ripasudil eye drops every 12 h (8 a.m., 8 p.m.) without discontinuing the currently used eye drops.

2.2. Parameters of Contact Lens Sensor Measurements

The SENSIMED Triggerfish® (Sensimed AG, Lausanne, Switzerland) comprises a wireless silicon CLS with embedded strain gauges, an antenna to be attached around the eye, and a recorder. A photograph of the CLS is presented in Figure 1. The CLS device can record IOP continuously for 24 h without disturbing the patient's daily routine [10]. The CLS records the IOP-related corneoscleral biodimensional change. However, the output is provided in millivolts equivalent (mVeq), which cannot be converted to IOP in millimeters of mercury (mmHg) directly. A highly positive correlation has been observed between the IOP measured using a CLS (mVeq) and that measured using a tonometer (mmHg) [11,12]. Otherwise, another study reported a low correlation between the IOP measured using a CLS (mVeq) and the IOP (mmHg) measured using the Tono-pen® XL applanation tonometer (r = 0.291) [13]. The value of IOP measured using a CLS (mVeq) could not be considered as the true IOP, but the fluctuation of the IOP value measured using a CLS is considered effective and safe for clinical use and has been used in some clinical projects pertaining to glaucoma research [14].

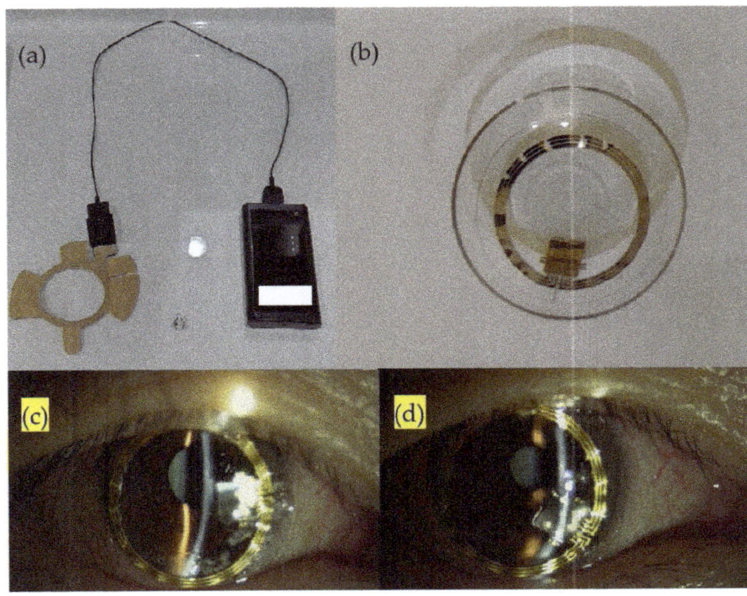

Figure 1. External photograph of CLS (**a**) antenna and recorder; (**b**) contact lens; (**c**) external eye photograph of the participant's ocular surface right after CLS administration; (**d**) external eye photograph after 24 h administration, mild conjunctiva congestion noted.

The CLS parameters included the average and SD values of IOP measurements in a 24 h day, at sleep and awake times; IOP fluctuations (difference between maximum value and minimum value) during the 24 h, at sleep and awake times; maximum value during the 24 h, at sleep and awake times; difference between the sleep and awake average IOPs; amplitude of the cosine fit curve; and IOP in mmHg measured using GAT. The CLS measured the IOP every 5 min; therefore, data (ID) were recorded 288 times within 24 h. The SD values, IOP fluctuations, maximum values for 24 h and sleep and awake times were calculated using RAW data. The circadian rhythm measurement was based on the cosine fit curve used to study the circadian biological rhythm. The cosine fit curve of the 24 h IOP measurements was examined using JMP statistical software with time-series analysis. The AUC was calculated using the integral of the cosine fit curve during sleep. To calculate the mean IOP of sleep time, the sleep time AUC was divided by the ID interval number of sleep time. To compare the circadian IOP fluctuations before and after the administration of ripasudil eye drops, the amplitudes of the cosine fit curve and the difference between the sleep time and awake time average IOPs were calculated. Ripasudil eye drops were administered at 8 a.m. and 8 p.m. According to Tanihara et al., the IOP-lowering effect of ripasudil persists for approximately 7 h after administration [9]. Therefore, in order to avoid the period that ripasudil might no longer be effective, we divided the awake time and sleep time IOP measurement period into two halves starting at 3 p.m. and 3 a.m., respectively, and SD values and IOP fluctuations of the first half in awake time and sleep time were calculated. A sample CLS measurement is presented in Figure 2.

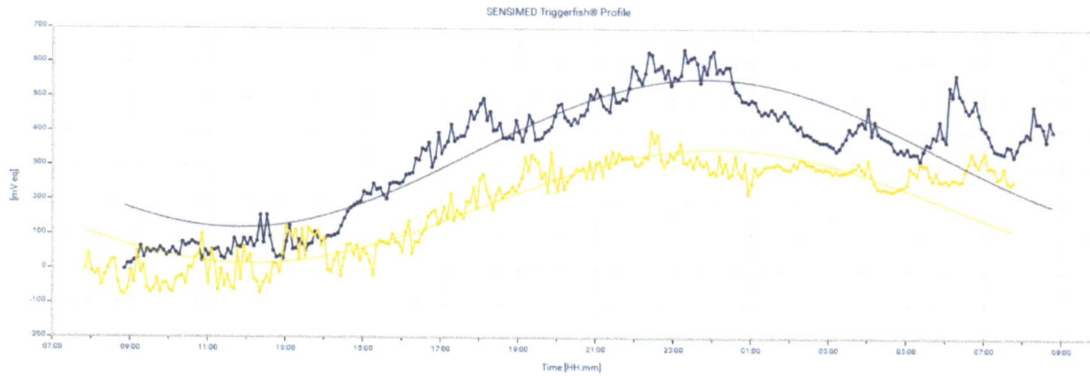

Figure 2. Example of the 24 h intraocular pressure-related pattern before and after ripasudil eye drop administration. Contact lens sensor (CLS) pattern before (blue tracing) and after (yellow tracing) administration. Smooth curve: cosine fit curve. Polygonal line: CLS raw data.

2.3. Statistical Analysis

The paired samples were compared using paired *t*-tests. Except where stated otherwise, the data are presented as mean ± SD values. All analyses were performed using JMP® PRO 15 (SAS institute, Inc., Cary, NC, USA 2019).

2.4. Ethical Approval

The ethics committee of the Yokohama City University Graduate School of Medicine approved the study, and written informed consent was obtained from all participants. The study adhered to the contents of the Declaration of Helsinki. The study was registered (UMIN000041093) in the University Hospital Medical Information Network Clinical Trials Registry (UMIN-CTR) of Japan.

3. Results

The study included one patient with POAG and five patients with NTG. The ocular characteristics and demographic features of the participants are summarized in Table 1. None of the patients withdrew their participation in the study due to the adverse effects of ripasudil. Transient conjunctival congestion and blurred vision were noted in all cases after the CLS examination; however, these symptoms improved within one day. No other adverse effects, such as corneal ulcers, were noted.

Table 1. Baseline demographic and ocular characteristics of the study participants.

Characteristic	Value
Age, years (mean ± SD)	58.33 ± 13.89
Sex (n)	
Male	4
Female	2
Ancestry (n)	
Asian	6
SER, D (mean ± SD)	−2.96 ± 1.6
CCT, µm (mean ± SD)	516.5 ± 28.85
IOP, mmHg (mean ± SD)	13.167 ± 2.714
Visual field parameters (mean ± SD)	
MD, dB	−4.78 ± 3.3
PSD, dB	6.56 ± 4.54
VFI, %	88.33 ± 8.31
OCT parameters	
RNFL thickness, µm (mean ± SD)	71.08 ± 9.36
Medication number (mean ± SD)	2 ± 1.09

Abbreviations: CCT, central corneal thickness; D, diopters; IOP, intraocular pressure; MD, mean deviation; OCT, optical coherence tomography; PSD, pattern standard deviation; RNFL, retinal nerve fiber layer; SD, standard deviation; SER, spherical equivalent refraction; VFI, visual field index.

The results of the paired t-tests before and after the administration of ripasudil are presented in Table 2. Although the value of each parameter became lower after ripasudil eye drop administration, the reduction did not reach statistical significance. The IOP measured in mmHg using GAT also did not reveal statistical significance.

Table 2. Comparison of the differences between intraocular pressure-related parameters before and after administering ripasudil eye drops.

CLS Parameter, GAT IOP	Before Ripasudil	After Ripasudil	p-Value
24 h average IOP, mVeq	218.760 ± 80.100	102.311 ± 97.784	0.0265 *
24 h IOP fluctuation, mVeq	433.467 ± 135.872	407.867 ± 90.102	0.3337
24 h IOP maximum, mVeq	407.300 ± 147.486	337.150 ± 91.250	0.1331
Sleep time average IOP, mVeq	296.774 ± 106.009	216.554 ± 110.489	0.0551
Sleep time average IOP before 3 a.m., mVeq	297.879 ± 125.821	221.780 ± 125.879	0.0855
Sleep time IOP fluctuation, mVeq	173.817 ± 70.431	146.133 ± 35.187	0.2529
Sleep time IOP fluctuation before 3 a.m., mVeq	166.250 ± 73.625	118.500 ± 28.613	0.1358
Sleep time IOP maximum value, mVeq	396.719 ± 148.986	291.650 ± 114.563	0.0723
Awake time average IOP, mVeq	178.478 ± 67.601	89.808 ± 71.734	0.0093 *
Awake time average IOP before 3 p.m., mVeq	94.717 ± 55.398	31.599 ± 121.857	0.1039
Awake time IOP fluctuation, mVeq	386.400 ± 153.716	370.67 ± 112.953	0.3828
Awake time IOP fluctuation before 3 p.m., mVeq	158.867 ± 44.401	142.033 ± 39.801	0.0647
Awake time IOP maximum value, mVeq	360.217 ± 160.586	299.300 ± 98.133	0.1439
24 h IOP SD, mVeq	105.442 ± 49.723	102.602 ± 36.562	0.4453
Sleep time IOP SD, mVeq	39.647 ± 16.677	32.808 ± 6.187	0.2306
Sleep time IOP SD before 3 a.m., mVeq	39.729 ± 20.588	25.244 ± 8.132	0.1250
Awake time IOP SD, mVeq	103.148 ± 60.864	94.902 ± 44.280	0.3716
Awake time IOP SD before 3 p.m., mVeq	36.601 ± 6.086	31.726 ± 14.034	0.1305
Difference between sleep time average IOP and awake time average IOP, mVeq	118.296 ± 46.641	126.745 ± 54.248	0.6553
Amplitude of cosine fit curve, mVeq	122.184 ± 67.493	116.234 ± 60.533	0.4136
Sleep time AUC of cosine fit curve/ID interval number	297.668 ± 112.092	205.650 ± 124.905	0.0570
GAT, mmHg	13.167 ± 2.714	12.333 ± 1.751	0.1446

* Statistically significant. Data are presented as mean ± SD. Abbreviations: AUC, area under the curve; CLS, contact lens sensor; GAT, Goldmann applanation tonometry; SD, standard deviation; IOP, intraocular pressure.

4. Discussion

The lowering of nocturnal IOP plays a crucial role in the daily treatment of glaucoma, and the nocturnal IOP-lowering effects of glaucoma medications have been reported in previous studies. β-blockers (e.g., timolol) are less effective in reducing IOP during nighttime hours [15–17]. The efficacy of carbonic anhydrase inhibitors (CAIs) in lowering nocturnal IOP remains controversial, as a significant reduction in nocturnal IOP was reported by some studies [18,19], whereas ineffectiveness in lowering IOP during the nocturnal hours was reported by another study [20]. β-blockers and CAIs lower IOP by suppressing aqueous humor production. A previous study revealed the presence of a circadian rhythm in aqueous humor production; the aqueous humor production is lower at night, and this might make the IOP-lowering effect of β-blockers and CAIs less effective at night [21]. Prostaglandin (PG) analogs lower IOP by affecting the uveoscleral outflow. Notably, among PG analogs, different medications possess varying nocturnal IOP-lowering effects, even with similar IOP-lowering mechanisms. Stewart et al. reported that for latanoprost, the mean reduction in nighttime IOP was significantly lower than that of daytime IOP. However, the same effect could not be noted for bimatoprost and travoprost [22]. Ripasudil reduces the IOP by changing the cell morphology of the TM. The IOP-lowering effect would presumably be less affected by the circadian rhythm of aqueous humor production. Therefore, theoretically, there might be a chance to reduce the nocturnal IOP and then reduce the IOP circadian fluctuation after ripasudil eye drop administration. However, the reduction in circadian fluctuations did not reach statistical significance after a 2-week ripasudil eye drop administration in the present study. The possible reason is that the nocturnal IOP reduction amount was not significantly larger than the diurnal IOP reduction after ripasudil eye drop administration. However, there is no study comparing the reduction in nighttime IOP with that of daytime IOP after ripasudil eye drop administration directly. Further evaluation is necessary.

Due to the effect of the loosening of the TM and the juxtacanalicular tissue, the mechanism of the IOP-lowering effect of ripasudil is considered similar to that of selective laser trabeculoplasty (SLT). Previous studies reported the relationship, comparisons, and treatment effects of ripasudil and SLT. Baba et al. reported increased success rates of SLT in patients with ripasudil-effective glaucoma [23]. Ono et al. reported that both adjuvant ripasudil and SLT reduced IOP significantly in patients with inadequately controlled glaucoma, and the percentage reduction in IOP between the two groups was not statistically significant [24]. Tojo et al. used a CLS to measure the changes in IOP before and after SLT in patients with NTG [25]. They reported that the reduction in circadian IOP fluctuation before and after SLT was not statistically significant ($p = 0.77$); however, the reduction in the nocturnal IOP fluctuation was significant (before SLT, 290 $\pm$ 86 mVeq vs. after SLT, 199 $\pm$ 31 mVeq; $p = 0.014$). In the present report, the reduction in circadian IOP fluctuation, sleep time IOP fluctuation, IOP fluctuation of the first half of sleep time and reduction in IOP measured by GAT were all not statistically significant. The baseline IOP and the characteristics of patients were not similar between the previous reports and the present study, so we could hardly compare the IOP-fluctuation-reducing effect between SLT and ripasudil directly. Whether there is similarity in IOP lowering-mechanisms in regard to IOP fluctuation changes needs further evaluation.

The amplitude of the cosine fit curve is the parameter which is considered the circadian IOP fluctuation, and the reduction in amplitude is considered as an important factor for glaucoma control. However, some controversial results have been reported. Hoban et al. reported a positive trend between higher amplitude and visual field progression without statistical significance ($p = 0.053$) [26]. Tojo et al. reported that the amplitude of the cosine fit curve did not show a positive correlation with a rapid progression visual field, but a positive correlation did exist with the nocturnal maximum value, nocturnal IOP fluctuation (the difference of nocturnal maximum and minimum value) and SD value. Moreover, the SD value of IOP was the CLS parameter which was most associated with the rapid progression of visual field changes (24 h IOP SD, $p = 0.0404$; diurnal IOP SD, $p = 0.0330$; nocturnal IOP SD, $p = 0.0027$) [27]. SD represents the amount of dispersion in a dataset. Therefore, the CLS SD indicates the IOP fluctuation and is different from circadian fluctuation. The lower the value of SD, the more stable the IOP is. The findings of this previous study could imply that a constant and stable IOP at any time in a day is important for glaucoma treatment. The SD of the 24 h IOP measured by the CLS in patients with NTG was greater than that of healthy individuals (patients with NTG, 112.51 $\pm$ 26.90 mVeq vs. healthy individuals, 85.18 $\pm$ 29.61 mVeq; $p = 0.002$). Moreover, the sleep-time IOP SD was significantly lower than the awake-time IOP SD in patients with NTG [28]. The baseline SD values of the present study were quite similar with those of a previous report. Moreover, a lower IOP with larger IOP SD during awake time and higher IOP with lower IOP SD during sleep time were observed both in the present and previous studies [27,28]. According to the Goldmann equation, episcleral venous pressure (EVP) variation may affect the IOP variation. Heavier daily physical movements and a wider range of blood pressure variation during awake time could explain the large awake-time IOP SD. However, other factors, especially those that could affect sleep-time EVP, remain unclear. The quality of sleep and sleeping behavior may be possible factors; therefore, further studies on how the sleep physiology and autonomic nerve system affect EVP may possibly clarify these aspects. In the present study, the SD values of all the IOP categories were reduced within a day following ripasudil administration; however, no statistically significant difference was noted. To the best of our knowledge, no report focusing on the change in the IOP SD value following any glaucoma treatment exists. For lowering the IOP SD, lowering the IOP might be important; moreover, we believe that different methods other than lowering IOP with updated glaucoma treatments exist for lowering the IOP SD value.

The 24 h average IOP mVeq ($p = 0.0265$) and awake-time average IOP mVeq ($p = 0.0093$) were significantly lower after ripasudil eye drop administration than those before the ripasudil eye drop administration. Although it was not significant, a trend of reduction in sleep-time average IOP mVeq could be noted ($p = 0.0551$). As mentioned above, mVeq is the measurement of biodimensional change between the corneoscleral surface. Whether mVeq could be presented as a real IOP measurement remains controversial. There are some possible reasons that result in the change of corneoscleral surface after ripasudil eye drop administration, such as IOP reduction (IOP lowering effect of ripasudil), conjunctiva hyperemia and increasing vessel density in the sclera (ocular surface events resulted from ripasudil eye drop administration), etc. However, we could not definitively determine the reason. Further evaluation, such as 24 h IOP mmHg measured by tonometry and anterior segment OCT examination, may help in solving this question.

The reproducibility of CLS remains an important issue. The previous study reported that the overall mean correlation r = 0.59 was noted between the two measurements in the same patients [29]. We repeated the CLS measurements after ripasudil eye drop administration in two patients. A sample of CLS measurement reproducibility is presented in Figure 3. High positive correlations were noted (mean Pearson's correlation r = 0.854). The difference in the value of the two measurements was calculated and presented in Table 3. The difference value percentage of the awake-time maximum, awake-time fluctuation, sleep-time maximum value, and sleep-time fluctuation between the two measurements were quite similar to the previous report (29.5%, 40.4%, 12.9% and 22.9%, respectively) [27]; therefore, we believed that the validity of the present study was similar to the previous report.

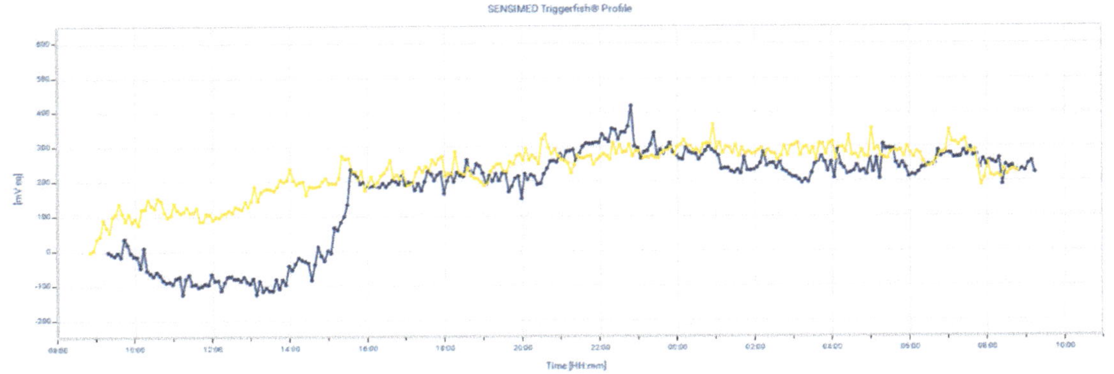

Figure 3. Example of the reproducibility of 24 h intraocular pressure after ripasudil eye drop administration using a CLS. r = 0.841. Blue tracing: first time measurement. Yellow tracing: second time measurement for reproducibility exam.

Table 3. The percent difference between the two CLS measurements after ripasudil eye drops administration in 2 cases.

CLS Parameter	Case Number 1	Case Number 3	Average
Awake-time maximum, mVeq	71.5/383.95 (18.6%)	134.3/335.55 (39.9%)	29.25%
Awake-time fluctuation, mVeq	194.8/445.6 (43.7%)	153.9/399.45 (38.5%)	41.1%
Sleep-time maximum, mVeq	24.6/352.5 (7%)	56.1/311.35 (18.0%)	12.5%
Sleep-time fluctuation, mVeq	37.1/125.35 (29.6%)	21/99.1 (21.2%)	25.4%

(The different values between the two measurements/the mean value of the two measurements) × 100%. Average percentage = (Case 1 percent difference + Case 3 percent difference)/2.

The correlation between CLS IOP and tonometry IOP remains controversial [11–13]. In order to measure the IOP (mmHg) in the supine position in the sleep time, we used an i care ic200 to measure the 24 h IOP fluctuation every 2 h in two patients. A sample of CLS IOP (mVeq) and ic200 IOP (mmHg) measurements in a participant after ripasudil eye drops administration is presented in Figure 4. The mean Pearson's correlation between CLS IOP and ic200 IOP was r = 0.464. However, we could not compare the correlation value with the results of the previous studies directly, due to the different tonometry methods used. The ratio of the circadian fluctuation of IOP measured by the ic200 and IOP measured by the CLS were calculated and presented in Table 4. The ratio between the two cases were similar.

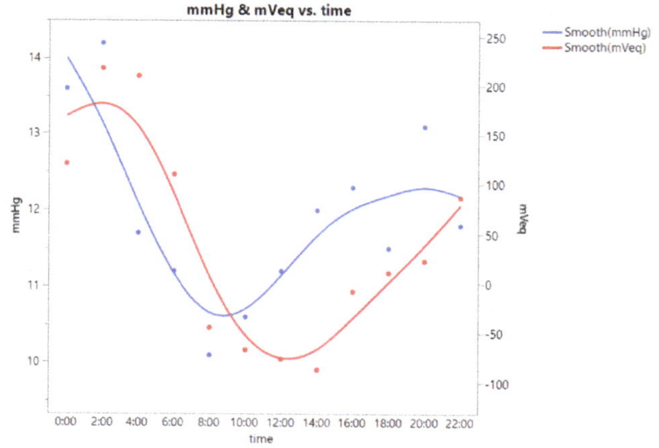

Figure 4. Example of the comparison of IOP (mmHg) using ic200 and IOP (mVeq) using CLS after ripasudil eye drop administration in a participant. r = 0.558. Blue smooth tracing: IOP (mmHg) using ic200. Red smooth tracing: IOP (mVeq) using CLS.

Table 4. The comparison between the CLS measurements and i care ic200 measurements after ripasudil eye drop administration in 2 cases.

	Case Number 1	Case Number 4
Pearson's correlation between CLS IOP and ic200 IOP	0.370	0.558
Circadian IOP fluctuation of ic200, mmHg	5.9	4.1
Amplitude of cosine fit curve of CLS, mVeq	184.13	109.52
Circadian IOP fluctuation of ic200/amplitude of CLS	0.032	0.037

In clinical performance, ripasudil eye drops are often used as a second-line treatment. Therefore, this study was intentionally conducted without discontinuing the currently used eye drops to replicate real-world clinical performance. Tanihara et al. reported increased IOP reduction with a high IOP at baseline [30]. Inazaki et al. reported an IOP reduction of −2.6 mmHg from baseline (baseline IOP: 17.9 ± 4.5 mmHg) after 12 months following treatment in patients with glaucoma receiving the maximum number of medications after adjunctive ripasudil eye drops administration [31]. In the present study, the baseline IOP was lower than those of the previous studies; therefore, a dampened IOP reduction could be expected. Moreover, the reduction in IOP measured by GAT did not reach statistical significance, nor did the IOP fluctuation measured by CLS. Further study is necessary to evaluate whether the attenuated IOP reduction may relate to attenuated IOP fluctuation reduction.

The present study had some limitations. First, owing to the high percentage of transient blurred vision and irritable sensations experienced while wearing the CLS and the high occurrence of conjunctival hyperemia following ripasudil eye drop administration, recruiting patients, especially older adult patients who had no experience of wearing contact lenses, was challenging. Moreover, since our study required the participants to receive CLS measurements twice, recruitment was even more challenging. Therefore, the number of participants was small. Second, a previous study reported that many physiological events, such as body position, blood pressure fluctuations, etc., could affect the relationship between IOP and ocular dimensional measurements [32]. We did not instruct the participants to sleep in a specific position or record the body position during sleeping; however, we had instructed the participants to sleep as they usually sleep in ordinary life to acquire the real IOP in their daily life. Third, the potentially unknown artifacts of CLSs, such as the upward drift phenomenon and contact lens fitting condition, were not considered. The upward drift phenomenon, which occurs after wearing the CLS overnight, results from a change in the corneal curvature [33,34]. How this phenomenon affects the accuracy of CLS measurements remains unclear. However, we conducted the study twice, and compared the data of the patients before and after ripasudil eye drop administration. Therefore, the factors that could possibly affect the validity of CLS, such as the contact lens–cornea fitting conditions, cornea edema, etc., should be consistent in these two measurements and be balanced out. Lastly, contact lenses may act as barriers that affect the absorption of eye drops and weaken the IOP-lowering effect. Therefore, the true effect of eye drops could be larger than that measured using the CLS in this study.

5. Conclusions

The reduction in circadian IOP fluctuation, nocturnal IOP fluctuation and SD value of IOP did not reach statistical significance as measured using a CLS after a 2-week adjunctive administration of ripasudil eye drops in patients with glaucoma. The baseline IOP was lower in the present study. The reduction in IOP measured by GAT also did not reach statistical significance. Further study is necessary to evaluate whether the low baseline IOP with less IOP reduction is the reason for the attenuated IOP fluctuation reduction.

Author Contributions: Conceptualization, S.-K.H.; formal analysis, S.-K.H.; investigation, S.-K.H.; methodology, M.I. and Y.M.; project administration, M.T. and T.K.; supervision, E.N.; writing—original draft preparation, S.-K.H.; writing—review and editing, N.M. All authors have read and agreed to the published version of the manuscript.

Funding: This research received no external funding.

Institutional Review Board Statement: The ethics committee of the Yokohama City University Graduate School of Medicine approved the study. The study adhered to the contents of the Declaration of Helsinki. The study was registered (UMIN000041093) in the University Hospital Medical Information Network Clinical Trials Registry (UMIN-CTR) of Japan.

Informed Consent Statement: Informed consent was obtained from all subjects involved in the study.

Data Availability Statement: The data are saved in archives which are managed and protected by the Department of Ophthalmology and Visual Science, Yokohama City University Graduate School of Medicine. In principle, the data are not publicly available.

Acknowledgments: The authors sincerely thank Yusuke Saigusa, Department of Biostatics, Yokohama City University School of Medicine, Yokohama, Kanagawa, Japan, for counseling us regarding the statistical analyses. The authors thank SEED™ for product introduction.

Conflicts of Interest: The authors declare no conflict of interest.

References

1. Weinreb, R.N.; Khaw, P.T. Primary Open-Angle Glaucoma. *Lancet* **2004**, *363*, 1711–1720. [CrossRef] [PubMed]
2. Mansouri, K.; Weinreb, R.N.; Medeiros, F.A. Is 24-h Intraocular Pressure Monitoring Necessary in Glaucoma? *Semin. Ophthalmol.* **2013**, *28*, 157–164. [CrossRef] [PubMed]
3. Liu, J.H.; Bouligny, R.P.; Kripke, D.F.; Weinreb, R.N. Nocturnal Elevation of Intraocular Pressure is Detectable in the Sitting Position. *Investig. Ophthalmol. Vis. Sci.* **2003**, *44*, 4439–4442. [CrossRef] [PubMed]

4. Liu, J.H.; Zhang, X.; Kripke, D.F.; Weinreb, R.N. Twenty-Four-Hour Intraocular Pressure Pattern Associated with Early Glaucomatous Changes. *Investig. Ophthalmol. Vis. Sci.* **2003**, *44*, 1586–1590. [CrossRef]
5. Arora, N.; McLaren, J.W.; Hodge, D.O.; Sit, A.J. Effect of Body Position on Epsicleral Venous Pressure in Healthy Subjects. *Investig. Ophthalmol. Vis. Sci.* **2017**, *58*, 5151–5156. [CrossRef] [PubMed]
6. Liu, J.H.; Weinreb, R.N. Monitoring Intraocular Pressure for 24 h. *Br. J. Ophthalmol.* **2011**, *95*, 599–600. [CrossRef]
7. Inoue, T.; Tanihara, H. Ripasudil Hydrochloride Hydrate: Targeting Rho Kinase in the Treatment of Glaucoma. *Expert Opin. Pharmacother.* **2017**, *18*, 1669–1673. [CrossRef]
8. Kusuhara, S.; Nakamura, M. Ripasudil Hydrochloride Hydrate in the Treatment of Glaucoma: Safety, Efficacy, and Patient Selection. *Clin. Ophthalmol.* **2020**, *14*, 1229–1236. [CrossRef]
9. Tanihara, H.; Inoue, T.; Yamamoto, T.; Kuwayama, Y.; Abe, H.; Suganami, H.; Araie, M. Intra-Ocular Pressure-Lowering Effects of a Rho Kinase Inhibitor, Ripasudil (K-115), over 24 h in Primary Open-Angle Glaucoma and Ocular Hypertension: A Randomized, Open-Label, Crossover Study. *Acta Ophthalmol.* **2015**, *93*, e254–e260. [CrossRef]
10. Lorenz, K.; Korb, C.; Herzog, N.; Vetter, J.M.; Elflein, H.; Keilani, M.M.; Pfeiffer, N. Tolerability of 24-h Intraocular Pressure Monitoring of a Pressure-Sensitive Contact Lens. *J. Glaucoma* **2013**, *22*, 311–316. [CrossRef]
11. Mansouri, K.; Shaarawy, T. Continuous Intraocular Pressure Monitoring with a Wireless Ocular Telemetry Sensor: Initial Clinical Experience in Patients with Open Angle Glaucoma. *Br. J. Ophthalmol.* **2011**, *95*, 627–629. [CrossRef]
12. Mansouri, K.; Weinreb, R.N.; Liu, J.H. Efficacy of a Contact Lens Sensor for Monitoring 24-h Intraocular Pressure Related Patterns. *PLoS ONE* **2015**, *10*, e0125530. [CrossRef] [PubMed]
13. Vitish-Sharma, P.; Acheson, A.G.; Stead, R.; Sharp, J.; Abbas, A.; Hovan, M.; Maxwell-Armstrong, C.; Guo, B.; King, A.J. Can the SENSIMED Triggerfish®lens data be used as an accurate measure of intraocular pressure? *Acta Ophthalmol.* **2018**, *96*, e242–e246. [CrossRef] [PubMed]
14. Dunbar, G.E.; Shen, B.Y.; Aref, A.A. The Sensimed Triggerfish Contact Lens Sensor: Efficacy, Safety, and Patient Perspectives. *Clin. Ophthalmol.* **2017**, *11*, 875–882. [CrossRef]
15. Konstas, A.G.P.; Mantziris, D.A.; Cate, E.A.; Stewart, W.C. Effect of Timolol on the Diurnal Intraocular Pressure in Exfoliation and Primary Open-Angle Glaucoma. *Arch. Ophthalmol.* **1997**, *115*, 975–979. [CrossRef]
16. Liu, J.H.; Kripke, D.F.; Weinreb, R.N. Comparison of the Nocturnal Effects of Once-Daily Timolol and Latanoprost on Intraocular Pressure. *Am. J. Ophthalmol.* **2004**, *138*, 389–395. [CrossRef]
17. Krag, S.; Andersen, H.B.; Sørensen, T. Circadian Intraocular Pressure Variation with Beta-Blockers. *Acta Ophthalmol. Scand.* **1999**, *77*, 500–503. [CrossRef]
18. Toris, C.B.; Zhan, G.L.; Yablonski, M.E.; Camras, C.B. Effects on Aqueous Flow of Dorzolamide Combined with Either Timolol or Acetazolamide. *J. Glaucoma* **2004**, *13*, 210–215. [CrossRef]
19. Bagga, H.; Liu, J.H.; Weinreb, R.N. Intraocular Pressure Measurements throughout the 24 h. *Curr. Opin. Ophthalmol.* **2009**, *20*, 79–83. [CrossRef]
20. Gulati, V.; Fan, S.; Zhao, M.; Maslonka, M.A.; Gangahar, C.; Toris, C.B. Diurnal and Nocturnal Variations in Aqueous Humor Dynamics of Patients with Ocular Hypertension Undergoing Medical Therapy. *Arch. Ophthalmol.* **2012**, *130*, 677–684. [CrossRef] [PubMed]
21. Maus, T.L.; McLaren, J.W.; Shepard, J.W., Jr.; Brubaker, R.F. The Effects of Sleep on Circulating Catecholamines and Aqueous Flow in Human Subjects. *Exp. Eye Res.* **1996**, *62*, 351–358. [CrossRef] [PubMed]
22. Stewart, W.C.; Konstas, A.G.; Nelson, L.A.; Kruft, B. Meta-analysis of 24-h Intraocular Pressure Studies Evaluating the Efficacy of Glaucoma Medicines. *Ophthalmology* **2008**, *115*, 1117–1122.e1. [CrossRef]
23. Baba, T.; Hirooka, K.; Nii, H.; Kiuchi, Y. Responsiveness to Ripasudil May Be a Potential Outcome Marker for Selective Laser Trabeculoplasty in Patients with Primary Open-Angle Glaucoma. *Sci. Rep.* **2021**, *11*, 5812. [CrossRef] [PubMed]
24. Ono, K.; Sakemi, F.; Marumoto, T. Intraocular Pressure-Lowering Effects of Ripasudil, a Rho-Kinase Inhibitor, and Selective Laser Trabeculoplasty as Adjuvant Therapy in Patients with Uncontrolled Glaucoma. *Int. Ophthalmol.* **2021**, *41*, 605–611. [CrossRef] [PubMed]
25. Tojo, N.; Oka, M.; Miyakoshi, A.; Ozaki, H.; Hayashi, A. Comparison of Fluctuations of Intraocular Pressure before and after Selective Laser Trabeculoplasty in Normal-Tension Glaucoma Patients. *J. Glaucoma* **2014**, *23*, e138–e143. [CrossRef]
26. Hoban, K.; Peden, R.; Megaw, R.; Halpin, P.; Tatham, A.J. 24-h Contact Lens Sensor Monitoring of Intraocular Pressure-Related Profiles in Normal-Tension Glaucoma and Rates of Disease Progression. *Ophthalmic Res.* **2017**, *57*, 208–215. [CrossRef]
27. Tojo, N.; Hayashi, A.; Otsuka, M. Correlation between 24-h Continuous Intraocular Pressure Measurement with a Contact Lens Sensor and Visual Field Progression. *Graefes Arch. Clin. Exp. Ophthalmol.* **2020**, *258*, 175–182. [CrossRef]
28. Kim, Y.W.; Kim, J.S.; Lee, S.Y.; Ha, A.; Lee, J.; Park, Y.J.; Kim, Y.K.; Jeoung, J.W.; Park, K.H. Twenty-Four-Hour Intraocular Pressure-Related Patterns from Contact Lens Sensors in Normal-Tension Glaucoma and Healthy Eyes: The Exploring Nyctohemeral Intraocular Pressure Related Pattern for Glaucoma Management (ENIGMA) Study. *Ophthalmology* **2020**, *127*, 1487–1497. [CrossRef]
29. Mansouri, K.; Medeiros, F.A.; Tafreshi, A.; Weinreb, R.N. Continuous 24-h monitoring of intraocular pressure patterns with a contact lens sensor: Safety, tolerability, and reproducibility in patients with glaucoma. *Arch. Ophthalmol.* **2012**, *130*, 1534–1539. [CrossRef]

30. Tanihara, H.; Kakuda, T.; Sano, T.; Kanno, T.; Imada, R.; Shingaki, W.; Gunji, R. Safety and Efficacy of Ripasudil in Japanese Patients with Glaucoma or Ocular Hypertension: 3-Month Interim Analysis of ROCK-J, a Post-Marketing Surveillance Study. *Adv. Ther.* **2019**, *36*, 333–343. [CrossRef]
31. Inazaki, H.; Kobayashi, S.; Anzai, Y.; Satoh, H.; Sato, S.; Inoue, M.; Yamane, S.; Kadonosono, K. One-Year Efficacy of Adjunctive Use of Ripasudil, a Rho-Kinase Inhibitor, in Patients with Glaucoma Inadequately Controlled with Maximum Medical Therapy. *Graefes Arch. Clin. Exp. Ophthalmol.* **2017**, *255*, 2009–2015. [CrossRef] [PubMed]
32. Lee, T.E.; Yoo, C.; Lin, S.C.; Kim, Y.Y. Effect of Different Head Positions in Lateral Decubitus Posture on Intraocular Pressure in Treated Patients with Open-Angle Glaucoma. *Am. J. Ophthalmol.* **2015**, *160*, 929–936.e4. [CrossRef] [PubMed]
33. Beltran-Agulló, L.; Buys, Y.M.; Jahan, F.; Shapiro, C.M.; Flanagan, J.G.; Cheng, J.; Trope, G.E. Twenty-Four Hour Intraocular Pressure Monitoring with the SENSIMED Triggerfish Contact Lens: Effect of Body Posture During Sleep. *Br. J. Ophthalmol.* **2017**, *101*, 1323–1328. [CrossRef] [PubMed]
34. Hubanova, R.; Aptel, F.; Chiquet, C.; Mottet, B.; Romanet, J.P. Effect of Overnight Wear of the Triggerfish(®) Sensor on Corneal Thickness Measured by Visante(®) Anterior Segment Optical Coherence Tomography. *Acta Ophthalmol.* **2014**, *92*, e119–e123. [CrossRef]

Journal of *Personalized Medicine*

MDPI

Article

Personalized Approach in Treatment of Neovascular Age-Related Macular Degeneration

Radina Kirkova [1,2,*], Snezhana Murgova [1,3], Vidin Kirkov [4] and Ivan Tanev [5]

1 Department of Ophthalmology, ENT and Maxillofacial Surgery, Medical University—Pleven, 5800 Pleven, Bulgaria
2 Department of Ophthalmology, IRCCS Humanitas Research Hospital, 20089 Rozzano, Italy
3 Department of Ophthalmology, University Hospital Dr. Georgi Stranski, 5800 Pleven, Bulgaria
4 Department of Health Policy and Management, Faculty of Public Health "Prof. Dr. Tzekomir Vodenicharov", Medical University of Sofia, 1000 Sofia, Bulgaria
5 Eye Clinic "Zrenie", 1000 Sofia, Bulgaria
* Correspondence: dr_rkirkova@abv.bg; Tel.: +35-9883333880

Abstract: Background: Age-related macular degeneration (AMD) is a progressive, degenerative disease of the central retina. AMD is subdivided into "dry" (atrophic), "wet" (exudative), and neovascular (nAMD) forms. In recent years, the concepts about nAMD changed with the development of optical coherence tomography–angiography (OCT-A) and intravitreal anti-VEGF treatment. The aim of this study was to define the morphologic type of the neovascular membrane (NVM) before treatment with OCT-A and to register vascular remodeling after treatment with anti-VEGF. We also analyzed the relationship between NVM and visual acuity. Methods: The study was retrospective and included 119 patients with newly diagnosed, treatment-naïve nAMD. All the patients underwent full ophthalmic examination and also fluoresceine angiography and optical coherence tomography–angiography (OCT-A). Results: Based on the collected data, we found repetitive regularities. Conclusion: The analysis of our results could be used as prognostic markers for the evolution of the disease and as a basis for new treatment strategies, depending on the naïve NVM morphologic type.

Keywords: OCT-A; neovascularization; AMD; anti-VEGF; vascular remodeling; personalized; approach

Citation: Kirkova, R.; Murgova, S.; Kirkov, V.; Tanev, I. Personalized Approach in Treatment of Neovascular Age-Related Macular Degeneration. *J. Pers. Med.* **2022**, *12*, 1456. https://doi.org/10.3390/jpm12091456

Academic Editor: Chieh-Chih Tsai

Received: 8 August 2022
Accepted: 1 September 2022
Published: 5 September 2022

Publisher's Note: MDPI stays neutral with regard to jurisdictional claims in published maps and institutional affiliations.

1. Introduction

AMD is a progressive, degenerative disease, involving the central retina. It was first described by Hutchinson in 1874 as a "symmetric chorioretinal disease in old people". Many years later, the term "age-related maculopathy" was introduced. Nowadays, the end stage of the disease is called AMD. The disease comprises 8.7% of the reasons for blindness worldwide [1–4] and it occupies third place, after cataracts and glaucoma [5]. According to the World Health Organization (WHO), in 2040, patients with AMD will exceed 288 million. AMD is subdivided into "dry" (atrophic), "wet" (exudative), and neovascular. All the forms result in irreversible vision loss and heavy impairment of the quality of life of the patients [5]. Before the introduction of anti-VEGF for ophthalmic purposes in 2006, nAMD was considered the worst because of the development of new vessels, which leads to exudation and bleeding in and under the neurosensory retina, resulting in fast vision loss. For the diagnosis of nAMD, we need ophthalmoscopy, but also highly specialized tests such as fluorescein angiography (FA), indocyanine green angiography (ICG), optical coherence tomography (OCT), and OCT-angiography (OCT-A). OCT and OCT-A were developed as devices for early diagnosis and registration of the stage and the size of the lesions in patients with AMD [6–8]. The use of OCT-A in everyday clinical practice leads us to the necessity of introducing new diagnostic parameters and classifications because of its unique features and specific relationships between imaging and histology, much more complex than those of the other imaging techniques available in the ophthalmology [9].

2. Aim and Scope

The aim of this study was to evaluate the abilities of OCT-A in the diagnosis and in the follow-up of the results after treatment with intravitreal anti-VEGF in patients with nAMD.

Aim 1: Defining the type of NVM before treatment (naïve), according to OCT-A images.

Aim 2: Assessment of the changes in the type of NVM after treatment with intravitreal anti-VEGF (vascular remodeling).

Aim 3: Assessment of the relationship between visual acuity (VA) before treatment and VA after treatment.

Aim 4: Assessment of the relationship between the type of NVM (before and after treatment) and VA (before and after treatment).

3. Materials and Methods

All materials were collected in the following clinics:

- Eye clinic "Zrenie"—Sofia, Bulgaria
- Eye clinic in University Hospital for Active Treatment "Georgi Stranski"—Pleven

Our study includes all the patients with newly diagnosed, treatment-naïve nAMD. The study was retrospective and included 119 patients (Table 1).

Table 1. Demographic characteristics of the followed cohort.

Age	Males	Females	Total
Mean age	75.75	75.80	75.78
SD	9.65	9.46	9.61
Median	77	77	77
Range	53	66	119

All the data were collected in the period November 2018–December 2021. The algorithm for diagnosis and evaluation of patients was clearly defined and equal for all patients with newly diagnosed, treatment-naïve nAMD. They underwent full ophthalmic examination:

- Full anamnesis—family history, risk factors, allergies, systemic diseases
- Visual acuity for distance and near through chart with Snellen optotipes. If the patient could not see optotipes, we examined their ability to count fingers and register the moving of a hand. In patients with very low visual function, we assessed their ability to detect light—perception and proection and if they can define the color of the light ("red" or "green").
- Tonometry ("air-puff" tonometer)
- Biomicroscopy—assessment of orbit, eyelids, lid margin, conjunctiva, cornea, iris, pupillary reactions, anterior chamber, lens (also LOCS grading), or IOL
- Indirect ophthalmoscopy—assessment of vitreous body, optic nerve, retina (ophthalmoscopic signs of exudative activity or drusen)

All the patients underwent FA ("gold standard"), structural OCT and OCT-A. Main scientific interest of this study is to assess vascular remodeling after anti-VEGF treatment through OCT-A. We used OCT-A machine Nidek RS-3000 Advance 2 (software version NAVIS-EX 1.8.0).

After performing all the tests described above and confirming the diagnosis, we proceed to a detailed evaluation of OCT-A images in their three different depths. In our research we included only eligible OCT-A scans with good quality of the image. All the scans were assessed by one experienced examinator. The type of neovascular membrane was determined by this examinator, according to the classification described above and presented by Coscas et al. [10]. The cohort of 119 patients was divided into several groups, based on the morphological appearance of the neovascular membrane from the OCT-A image.

Patients with:

1. NV-membrane type "Sea fan"—with eccentric feeding vessel, massive trunks with thin capillary ramifications (Figure 1A)
2. NV-membrane type "Medusa"—has a massive feeding vessel with centrifugal vascular trunks with thin capillaries (Figure 1B)
3. NV-membrane type "Dead tree"—has a massive main trunk and ramifications varying in size and caliber (Figure 1C)
4. NV-membrane type "Lace"—highly anastomotic vascular network, without main vessel (Figure 1D)
5. NV-membrane type "Filaments"—composed of many intertwining, filamentous vessels (Figure 1E)
6. nondetermined type of NV-membrane—when the OCT-A appearance does not correspond to any of the above types

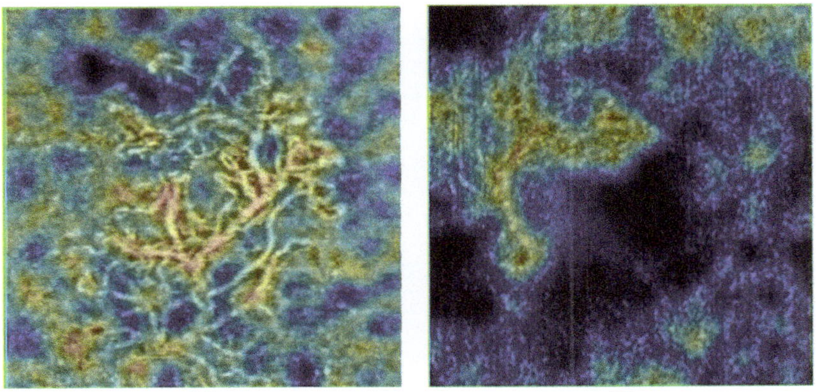

(A) **(B)**

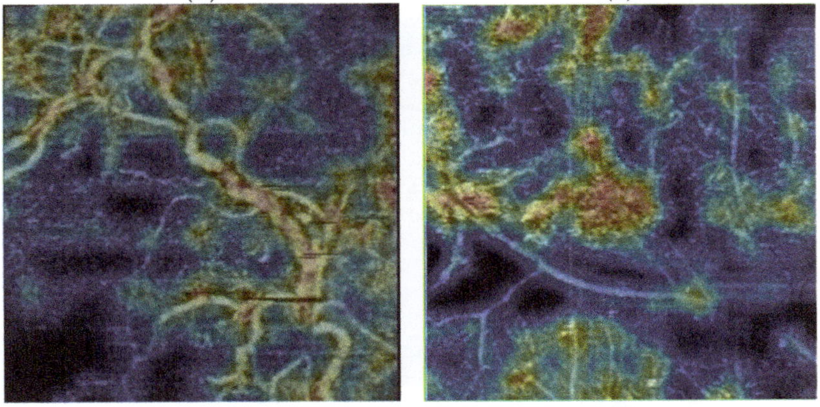

(C) **(D)**

Figure 1. *Cont.*

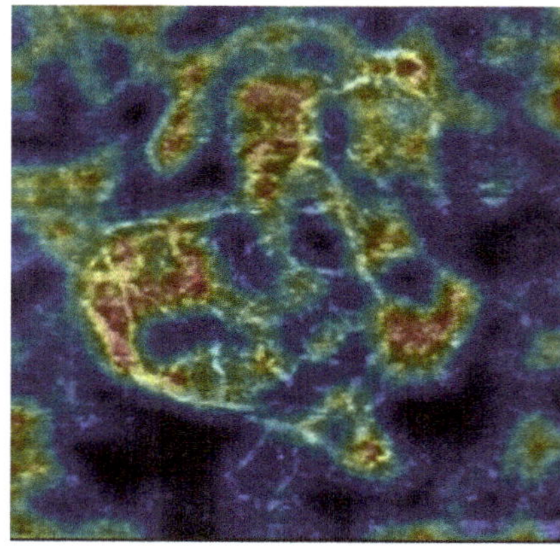

(E)

Figure 1. C-angiography, Eye Clinic "Zrenie"; Nidek OCT-A RS-3000) Advance 2. (**A**) NV type "Sea fan"; (**B**)NV type "Medusa"; (**C**)NV type "Dead tree"; (**D**). NV type "Lace"; (**E**) NV type "Filaments".

Of the approved anti-VEGF drugs in Bulgaria, Eylea (Bayer) is the most affordable, as it is reimbursed by the National Health Insurance. Our aim was not to compare the therapeutic effect of the different anti-VEGF drugs, available on the market, but to demonstrate the potential of the OCT-A in diagnosis and personalization of the therapeutic protocol, resulting in better structural and visual outcome. Therefore, only patients treated with the same drug were included in the study.

The anti-VEGF administration procedure is performed in an operating room under sterile conditions. Seven days before the intravitreal injection, the patient puts antibiotic drops for prophylaxis. After the placement of blepharostat, we perform anesthesia with Alcaine drops and double exposure to Betadine solution for 2 min, intrapalpebral. Using a caliper and depending on the patient's phakic status, Eylea (Bayer) is injected intravitreally into the lower nasal quadrant of the eyeball. The patient is bandaged until the next day. Control examinations, including assessment of visual acuity, tonometry, biomicroscopy, ophthalmoscopy, OCT, and OCT angiography, were performed on day 25 after the injection. Regarding admission to the current study of the therapeutic protocol, the next injection should be given after 1 month from the start, with the patient facing three "loading" doses one month apart. After each intravitreal injection, special attention must be paid to visual acuity as a functional measure of the effect of the treatment. In OCT-A images, a change in the morphological shape of the membrane is determined. Signs of progression are monitored—persistence, increase, or regression.

4. Statistical Analysis

In order to make the follow-up easier and more representative, the collected data were summarized and entered in a tabular form in Microsoft Excel. Each individual patient was introduced on a new line, and the indicators of morphological appearance of the naïve NV-membrane (NVM), morphological appearance of the NVM after three intravitreal applications of anti-VEGF, assessment of signs of progression and visual acuity before and after therapy were introduced in columns. Statistical analysis was performed using the software package SPSS, 13.0 (SPSS Inc., Chicago, IL, USA). All values with $p < 0.05$ were considered statistically significant.

5. Results

According to Aim 1, the individual types of NV membranes were defined before starting therapy according to their OCT-A characteristics. The largest number was of the "sea fan" type (40 patients), followed by "lace" (25 patients), "filaments" (23 patients), and "Medusa" (19 patients). The obtained results are represented graphically in Figure 2.

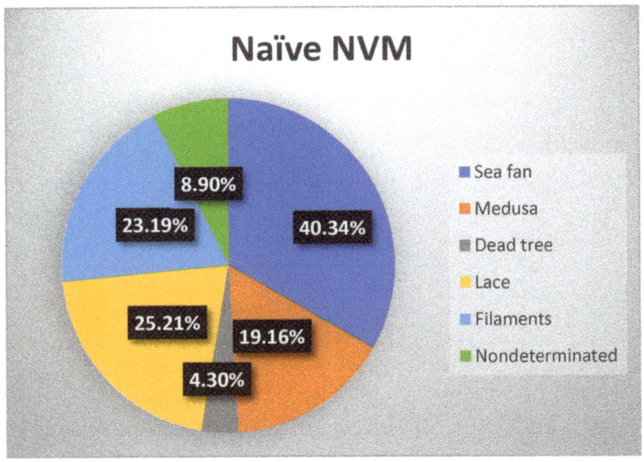

Figure 2. Percentage of naïve morphological type of NVM.

Although age is the major risk factor for the development and progression of the disease, no statistical relationship has been established between the age of the patients and the naïve form of the neovascular membrane. No relationship was found between the sex of the patients and the shape of the "naïve" neovascular membrane.

Aim 2: to determine whether there is a relationship between the initial appearance (naïve neovascular membrane) and its change after three applications of anti-VEGF intravitreal in a month. Statistical analysis proved a relationship between the morphologic form before therapy and that after for the whole cohort of 119 patients: Spearman's rho: $\rho = 0.61137, p = 0, n = 119$. This relationship is graphical represented in Figure 3.

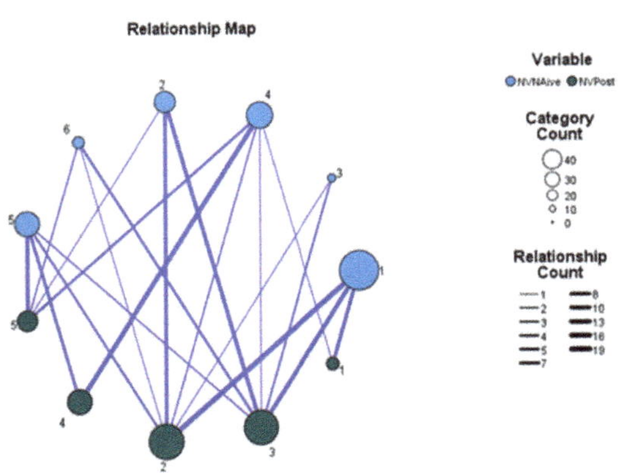

Figure 3. Relationship map: between the naïve type of NV membrane and the type after therapy: The thickness of the line is directly proportional to the strength of the connection. Legend: NV naïve—"naive" form of a neovascular membrane, before therapy; NV post—a form of the neovascular membrane after three intravitreal anti-VEGF applications with an interval one month. 1—Sea fan; 2—Medusa; 3—Dead tree; 4—Lace; 5—filaments; 6—undetermined.

For a more detailed and in-depth analysis, naive neovascular membranes were further subdivided according to the "maturity" of the vessels. This way, we seek for connection between the histological structure of the membrane (the components that make up its vessel wall) and the subsequent remodeling after therapy. The subdivision was performed taking into account the OCT-A image and caliber of the vessels forming the membrane and according to the theory of retinal angiogenesis of De Almodovar et al. [11].

Thus, two groups were formed: "differentiated" and "undifferentiated" membranes. The "differentiated" include "sea fan", "Medusa", "dead tree". To the group of "undifferentiated" we refer—"lace", "filaments", "undetermined". For the group of differentiated ($n = 62$), the Spearman's rho test was applied again, which confirmed the statistical correlation between the three morphological species: Spearman's rho: $\rho = 0.32385$, $p = 0.01024$, $n = 62$. A chi-square test was also applied—39.6406, $p = < 0.00001$, which confirmed once again the statistical relationship (the results are visible in Table 2).

Table 2. Results of the chi-square test.

	Results		
	NV Naïve	NV after Treatment	All Rows
Sea fan	40 (24.00) [10.67]	8 (24.00) [10.67]	48
Medusa	18 (23.00) [1.09]	28 (23.00) [1.09]	46
Dead tree	4 (15.00) [8.07]	26 (15.00) [8.07]	30
All columns	62	62	**124** (Total)

The study found that neovascular membranes of the "sea fan" type most often after therapy became "Medusa" (47%) or "dead tree" (33%). Only 20% did not change after therapy. "Medusa" type after therapy in 55.6% passed into a "dead tree", and in 44.4% did not change. In 75% of cases, the "dead tree" retained its morphological appearance after therapy. It is a final, maximally differentiated form (histologically) of a neovascular membrane, that does not change over time. The results are represented on Figure 4.

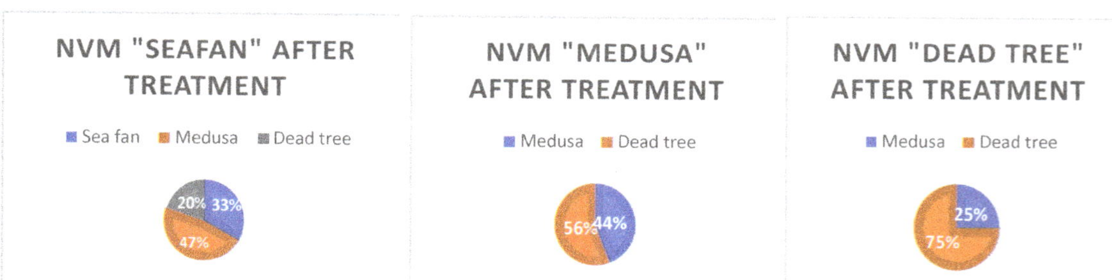

Figure 4. Remodeling NV type "sea fan", "Medusa", "dead tree" after therapy.

For the group of "undifferentiated", immature neovascular membranes, which include "lace", "filaments" and "nondetermined", Spearman's rho and chi-square test were applied again. However, they did not find a correlation between the different forms before and after therapy. Although in 64% of cases the naive neovascular membrane type "lace" does not change after therapy, the statistical data for the three types do not allow defining a relationship between naïve forms and the morphologic type of the membrane posttreatment.

According to requirement of Aim 3, we had to check if there is a link between visual acuity before and visual acuity after therapy. For the entire cohort of 119 patients and then for the two additionally subdivided groups (of "differentiated" and "undifferentiated" neovascular membranes), the statistical analysis showed an extremely strong relationship: "The higher the initial visual acuity, the higher it is after therapy".

In searching for the results of the Aim 4, an attempt was made to establish a connection between the individual morphological variants of naïve neovascular membranes and visual acuity. However, this has not been proven statistically. i.e., the data from the study did

not prove a relationship (either positive or negative) of any naïve morphological type of membrane with visual acuity before and after therapy.

After the first three "loading" doses of anti-VEGF and the registered on OCT-A vascular remodeling, a statistical relationship was established between the morphological variant after therapy and visual acuity for each group

For the group of "differentiated" neovascular membranes ("sea fan", "Medusa", "dead tree") the dependence is negative:

Spearman's rho: $\rho = -0.29387$, $p = 0.02043$, $n = 62$.

This means that the less differentiated the membrane (for example, "sea fan"), the lower the visual acuity.

For the group of undifferentiated membranes, again we have a statistically significant result, but it has a positive sign:

Spearman's rho: $\rho = 0.30498$, $p = 0.02228$, $n = 56$, i.e., the more differentiated the membrane (e.g., "filaments"), the higher the visual acuity.

6. Discussion

OCT-A was first demonstrated in 1997 by Chinn et al. [12] but they did not find applications in clinical practice—the quality of images was low due to a lack of technological capabilities. In the next decade, the technology developed and in 2006 Makita et al. presented the first OCT-A. It is based on a new method for analysis of high-resolution imaging techniques without the need to introduce contrast. OCT-A is based on the concept that in static eye tissues, the only moving structures are the formed elements of the blood as they pass through the vessels [13]. In clinical practice, OCT-A allows us to represent classical and occult neovascular membranes and gives detailed information about the choriocapillaris. OCT-A presents the retinal blood flow divided into two plexuses [14]: superficial and deep (formed by the complex internal/external plexus by morphological data). OCT-A allows us to study and to evaluate separately the two vascular plexuses and their demonstration in vivo. In fluorescein angiography (FA), the two plexuses overlap and cannot be evaluated separately, as the resulting images are cumulative. The possibility of visualization of the two plexuses separately is one of the main advantages of OCT-A. OCT-A demonstrates precisely the intravascular blood flow without contrast. When analyzing the image from the OCT-A, attention should be paid to the localization and to the depth of the scan with characteristics such as reflectivity, flow, morphology, and architectonic. We evaluate the four depths of scanning [15]: superficial plexus, deep plexus, RPE/Bruch membrane complex, and choroidal. The correct interpretation of the OCT-angiography image requires consideration of "en face" and transversal scans at each of the specified levels so that the clinical correlation can be maximally correct. For each of the four depths of scanning should be determined: flow and decorrelation, vascular morphology and architectonics, texture—all up to one new indicator characterizing OCT-A and differing it from FA [16]. There is a well-known classification of neovascular membrane types (NVMs) according to the findings in the FA images. Type I NV (occult) is the most common: it starts from the choroid and reaches the Bruch membrane and retinal pigment epithelium (RPE). Type II NV is located subretinally. Type III—intraretinal (retinal angiomatous proliferations—RAP). Although the FA has been the gold standard in the diagnosis of neovascular AMD for more than 50 years, the classifications used for it are not applicable to OCT-A. Palejwala et al. was the first to report the application of OCT-A for early detection of type I NV. El Ameen et al. described type II NV using OCT-A. Other studies reported the administration of OCT-A in the detection of type III NV (RAP). The introduction of the wide use of OCT-A into daily clinical practice led to the necessity of new diagnostic parameters and new classifications. Coscas et al. in 2015 examined with OCT-A 80 eyes with exudative AMD and determined different morphological variants of NV. According to Coscas et al. and other researchers [15] of recent years, based on the images from OCT-A, NV membranes morphologically are defined as "seaf an", "Medusa", "dead tree", "lace", and "filaments" [17,18].

The results of ours study proved a link between the "naïve" morphological forms of neovascular membranes and defined a dependency on switching one naïve form to another after therapy—the so-called vascular remodeling. For more accurate and thorough interpretation and taking into account De Almodovar et al.'s theory of vascular genesis, the "naïve" neovascular membranes were further subdivided into groups depending on the caliber of the vessels building them (the histological structure, respectively). The larger the caliber of the vessels building the NV membrane, the "more mature" in histological terms they are. The study found that "naïve" NV membranes from the group "differentiated" ("sea fan", "Medusa", "dead tree") tend to pass into each other after therapy – graphically represented on Figure 5.

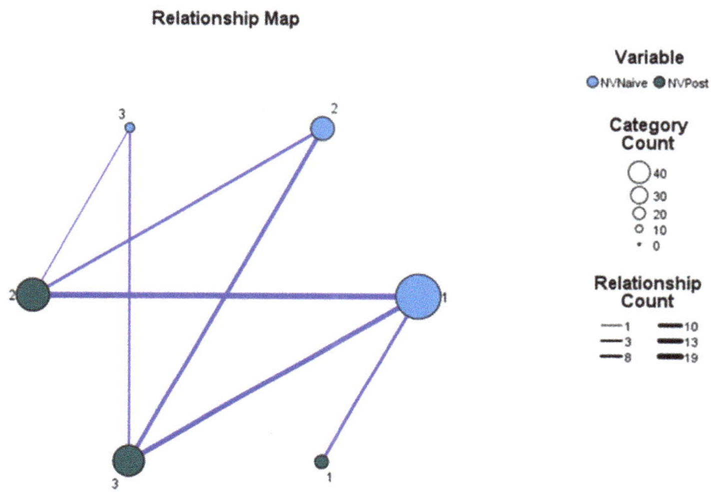

Figure 5. Relationship map—between the naïve type of NV membrane and the type after therapy—group "differentiated". The thickness of the line is directly proportional to the strength of the connection. Legend: NV naïve—"naïve" form of a neovascular membrane, before therapy; NV post—A form of the nonvascular membrane after three. intravitreal anti-VEGF applications with a month interval between them. 1—Sea fan; 2—Medusa; 3—Dead tree.

Another important regularity is the relationship of histological differentiation with visual acuity after therapy. It was found that the more differentiated a membrane (for example, "dead tree"), the higher is the visual acuity after treatment. For less differentiated membranes, no relationship for vessel remodeling has been proven. The relationship of the remodeled type of neovascular membrane (group "non-differentiated": "lace", "filaments", "nondetermined") was established after therapy with visual acuity. The more highly differentiated, and respectively histologically "mature" the type of the membrane, the higher the visual acuity. Given the tendency in highly differentiated naïve forms of neovascular membranes to pass into each other and their relationship to visual acuity after therapy, it can be concluded that when establishing a naïve form "sea fan", "Medusa", "dead tree", it is possible to predict the evolution of the disease and visual acuity after therapy.

7. Conclusions

Age-related macular degeneration increases its social importance because of the aging pyramid of the population. In its late stages, the disease is highly debilitating, disrupting the quality of life of patients and their ability to take full care of themselves. The neovascular form of AMD affects many layers of the outer and inner retina, the RPE, and the choroid. Multimodal imaging techniques (FA, Indocyanine angiography, OCT, OCT-angiography) are used in the diagnosis and follow-up of patients. Due to the possibilities for segment separation and analysis of microcirculation at different levels, OCT-A has become an integral part of the diagnosis and follow-up of patients with AMD. OCT-A improves the identification capabilities and complements structural OCT, FA, and indocyanine angiography. It is non-invasive, fast, and brings the ophthalmologist detailed information. The detected patterns for vascular wall remodeling could serve as prognostic markers for the evolution of the disease and to become a basis for developing new therapeutic regimens and individualizing the approach to the "naïve" type of NV membrane at the time of diagnosis of neovascular AMD.

Author Contributions: Conceptualization, R.K.; data curation, R.K.; formal analysis, R.K.; Investigation, R.K.; methodology, R.K. and V.K.; project administration, R.K.; resources, R.K.; software, V.K.; supervision, S.M. and I.T.; Validation, R.K.; visualization, R.K.; writing—original draft, R.K.; Writing—review & editing, R.K. All the four authors have equal contribution to the research. All authors have read and agreed to the published version of the manuscript.

Funding: This research received no external funding.

Institutional Review Board Statement: The study was approved by Ethic Committee of Eye Clinic "ZRENIE", Approval code: Z04/20.02.2018, Approval date: 20 February 2018; All procedures performed in studies involving human participants were in accordance with the ethical standards of the (Eye Clinic ZRENIE, Sofia, Bulgaria) and with the 1964 Helsinki declaration and its later amendments or comparable ethical standards. The study was reviewed by the Ethical Committee of Eye Clinic "Zrenie"—Sofia, Bulgaria.

Informed Consent Statement: Informed consent was obtained from all individual participants included in the study. All the three authors declare their consent to publish the research "Personalized approach in treatment of neovascular age-related macular degeneration".

Data Availability Statement: All the data and materials are saved in the archives of Eye Clinic "Zrenie": Sofia, Bulgaria. The information about patient's demographic characteristics, visual acuity (as a comment in each patient's folder) and OCT-A scans are saved in Nidek RS 3000 Advance (software version NAVIS-EX 1.8.0.) in Eye Clinic "Zrenie", Sofia, Bulgaria.

Conflicts of Interest: The authors declare no conflict of interest.

References

1. Mariotti, S.P.; Pascolini, D. Global estimates of visual impairment: 2010. *Br. J. Ophthalmol.* **2012**, *96*, 614–618.
2. Klein, R.; Lee, K.E.; Gangnon, R.E.; Klein, B.E. Incidence of visual impairment over a 20-year period: The Beaver Dam Eye Study. *Ophthalmology* **2013**, *120*, 1210–1219. [CrossRef] [PubMed]
3. Wang, J.J.; Rochtchina, E.; Lee, A.J.; Chia, E.-M.; Smith, W.; Cumming, R.G.; Mitchell, P. Tenyear incidence and progression of age-related maculopathy: The Blue Mountains Eye Study. *Ophthalmology* **2007**, *114*, 92–98. [CrossRef] [PubMed]
4. Weiner, D.E.; Tighiouart, H.; Reynolds, R.; Seddon, J.M. Kidney function, albuminuria and age-related macular degeneration in NHANES III. *Nephrol. Dial. Transplant.* **2011**, *26*, 3159–3165. [CrossRef] [PubMed]
5. Leske, M.C.; Wu, S.-Y.; Hennis, A.; Nemesure, B.; Yang, L.; Hyman, L.; Schachat, A.P. Nine-year incidence of age-related macular degeneration in the Barbados Eye Studies. *Ophthalmology* **2006**, *113*, 29–35. [CrossRef] [PubMed]
6. Gess, A.J.; Fung, A.E.; Rodriguez, J.G. Imaging in neovascular age-related macular degeneration. *Semin. Ophthalmol.* **2011**, *26*, 225–233. [CrossRef] [PubMed]
7. Adhi, M.; Duker, J.S. Optical coherence tomography—Current and future applications. *Curr. Opin. Ophthalmol.* **2013**, *24*, 213–221. [CrossRef]
8. Spaide, R.F.; Koizumi, H.; Pozzoni, M.C. Enhanced depth imaging spectral-domain optical coherence tomography. *Am. J. Ophthalmol.* **2008**, *146*, 496–500. [CrossRef]
9. Anger, E.M.; Unterhuber, A.; Hermann, B.; Sattmann, H.; Schubert, C.; Morgan, J.E.; Cowey, A.; Ahnelt, P.K.; Drexler, W. Ultrahigh resolution optical coherence tomography of the monkey fovea. Identification of retinal sublayers by correlation with semithin histology sections. *Exp. Eye Res.* **2004**, *78*, 1117–1125. [CrossRef] [PubMed]
10. Coscas, G.J.; Lupidi, M.; Coscas, F.; Cagini, C.; Souied, E.H. Optical coherence tomography angiography versus traditional multimodal imaging in assessing the activity of exudative age-related macular degeneration. A new diagnostic challenge. *Retina* **2015**, *35*, 2219–2228. [CrossRef] [PubMed]
11. de Almodovar, C.R.; Ny, A.; Carmeliet, P.; King, G.L.; Suzuma, K.; Sun, J.K.; Agostini, H.; Martin, G. Retinal Angiogenesis and Growth Factors. In *Retinal Vascular Disease*; Springer: Berlin/Heidelberg, Germany, 2007; pp. 38–77.
12. Enfield, J.; Jonathan, E.; Leahy, M. In vivo imaging of the microcirculation of the volar forearm using correlation mapping optical coherence tomography (cmOCT). *Biomed. Opt. Express* **2011**, *2*, 1184–1193. [CrossRef] [PubMed]

13. Potsaid, B.; Baumann, B.; Huang, D.; Barry, S.; Cable, A.E.; Schuman, J.S.; Duker, J.S.; Fujimoto, J.G. Ultrahigh speed 1050 nm swept source/Fourier domain OCT retinal and anterior segment imaging at 100,000 to 400,000 axial scans per second. *Opt. Express* **2010**, *18*, 20029–20048. [CrossRef] [PubMed]
14. Vakoc, B.J.; Lanning, R.M.; A Tyrrell, J.; Padera, T.; A Bartlett, L.; Stylianopoulos, T.; Munn, L.L.; Tearney, G.J.; Fukumura, D.; Jain, R.K.; et al. Three-dimensional microscopy of the tumor microenvironment in vivo using optical frequency domain imaging. *Nat. Med.* **2009**, *15*, 1219–1223. [CrossRef] [PubMed]
15. Souied, E.H.; El Ameen, A.; Semoun, O.; Miere, A.; Querques, G.; Cohen, S.Y. Optical coherence tomography angiography of type 2 neovascularization in age-related macular degeneration. *Dev. Ophthalmol.* **2016**, *56*, 52–56. [PubMed]
16. Farecki, M.L.; Gutfleisch, M.; Faatz, H.; Rothaus, K.; Heimes, B.; Spital, G.; Lommatzsch, A.; Pauleikhoff, D. Characteristiscs of type 1 and 2 VNV in exudative AMD in OCT Angiography. *Graefes Arch. Clin. Exp. Ophtahlmol.* **2017**, *255*, 913–921. [CrossRef] [PubMed]
17. Kuehlewein, L.; Bansal, M.; Lenis, T.L.; Iafe, N.A.; Sadda, S.R.; Filho, M.A.B.; De Carlo, T.E.; Waheed, N.K.; Duker, J.S.; Sarraf, D. Optical coherence tomography angiography of type 1 neovascularization in age-related macular degeneration. *Am. J. Ophthalmol.* **2015**, *160*, 739–748. [CrossRef] [PubMed]
18. Miere, A.; Semoun, O.; Cohen, S.Y.; El Ameen, A.; Srour, M.; Jung, C.; Oubraham, H.; Querques, G.; Souied, E.H. Optical coherence tomography angiography features of subretinal fibrosis in age-related macular degeneration. *Retina* **2015**, *35*, 2275–2284. [CrossRef] [PubMed]

Journal of
Personalized Medicine

MDPI

Article

Increased Orbital Muscle Fraction Diagnosed by Semi-Automatic Volumetry: A Risk Factor for Severe Visual Impairment with Excellent Response to Surgical Decompression in Graves' Orbitopathy

Christine Steiert [1,*], Sebastian Kuechlin [2], Waseem Masalha [1], Juergen Beck [1], Wolf Alexander Lagrèze [2] and Juergen Grauvogel [1]

1 Department of Neurosurgery, Medical Center—University of Freiburg, Faculty of Medicine, University of Freiburg, 79106 Freiburg, Germany; waseem.masalha@uniklinik-freiburg.de (W.M.); j.beck@uniklinik-freiburg.de (J.B.); juergen.grauvogel@uniklinik-freiburg.de (J.G.)
2 Eye Center, Medical Center—University of Freiburg, Faculty of Medicine, University of Freiburg, 79106 Freiburg, Germany; sebastian.kuechlin@uniklinik-freiburg.de (S.K.); wolf.lagreze@uniklinik-freiburg.de (W.A.L.)
* Correspondence: christine.steiert@uniklinik-freiburg.de; Tel.: +49-761-270-50010; Fax: +49-761-270-50080

Abstract: Graves' orbitopathy (GO) leads to increased orbital tissue and causes symptoms such as exophthalmos, functional complaints, or dysthyroid optic neuropathy. Different GO types with fat and/or muscle enlargement were identified, and increased muscle appears to particularly influence visual status and treatment response. The current study examines visual parameters dependent on orbital muscle volume fraction in a surgically treated GO cohort. After volumetric analysis of the preoperative orbital content, 83 orbits in 47 patients were categorized into predefined groups (increased or not-increased muscle fraction). All cases underwent pterional orbital decompression, and pre- and postoperative visual status was retrospectively analyzed. Forty-one orbits revealed increased and 42 orbits revealed not-increased muscle volume (mean fraction 29.63% versus (vs.) 15.60%). The preoperative visual acuity (VA) was significantly lower in orbits with increased vs. not-increased muscle volume (mean VA 0.30 vs. 0.53, difference 2.5 lines). After surgery, mean VA improved significantly by 1.7 lines in orbits with increased muscle volume. Not preoperative, but postoperative exophthalmos was significantly lower in orbits with not-increased muscle volume. Increased orbital muscle is associated with significantly reduced VA, but can be remarkably improved by pterional orbital decompression. Therefore, surgical therapy should be considered particularly in decreased VA with orbital muscle enlargement.

Keywords: orbital muscle volume; increased orbital muscle; orbital muscle enlargement; orbit volumetry; Graves' orbitopathy; Graves' disease; dysthyroid optic neuropathy; orbital decompression

Citation: Steiert, C.; Kuechlin, S.; Masalha, W.; Beck, J.; Lagrèze, W.A.; Grauvogel, J. Increased Orbital Muscle Fraction Diagnosed by Semi-Automatic Volumetry: A Risk Factor for Severe Visual Impairment with Excellent Response to Surgical Decompression in Graves' Orbitopathy. *J. Pers. Med.* **2022**, *12*, 937. https://doi.org/10.3390/jpm12060937

Academic Editor: Chieh-Chih Tsai

Received: 1 May 2022
Accepted: 30 May 2022
Published: 6 June 2022

1. Introduction

In Graves' orbitopathy (GO), a volume increase of orbital fat and/or muscle tissue due to an autoimmune inflammatory process leads to problems such as exophthalmos, double vision, and several functional complaints such as edema or ocular dryness [1]. The most severe complication is dysthyroid optic neuropathy (DON), where optic nerve compression causes optic nerve head swelling with the risk of consequent visual loss [2]. The therapeutic approach depends on the activity and severity of the disease [3]. Particularly in severe cases with DON, surgical decompression may be required [3,4]. Since prognosis in DON improves with early diagnosis and therapy induction, different clinical and imaging parameters have been studied to predict the risk of DON [5,6]. Imaging parameters focus on the analyses of the orbital content (e.g., the ratio of fat or muscle enlargement, the water fraction of the orbital tissue, or the content of the orbital apex ("apical crowding")) [6–10]. Different types of GO were identified depending on the increase in orbital muscle volume and/or fat volume [1,11]. Increased orbital apex crowding has been found to be associated with DON, but overall increased orbital muscle volume does not seem to necessarily be related to DON [5,6,12,13]. However, orbital muscle volume appears to be of central importance, and respective studies have focused on predictors of DON [12,14].

To the best of our knowledge, there has been no investigation of the possible influence of increased orbital muscle volume on ophthalmological functional parameters or their development after orbital decompression. The present study is the first to determine the preoperative orbital muscle volume in a GO cohort, which were all treated subsequently in a standardized fashion (pterional orbital decompression), and to evaluate differences in preoperative ophthalmological status and postoperative ophthalmological outcome dependent on an "increased" or "not-increased" orbital muscle volume (following Regensburg et al. [11]).

2. Patients and Methods

2.1. Study Design and Patient Population

The inclusion criterion was pterional orbital decompression between January 2001 and May 2021 due to GO with the availability of preoperative cranial imaging suitable for volumetry (computed tomography (CT) (soft tissue window, slice thickness $\leq$ 2.0 mm) or magnetic resonance imaging (MRI) (three dimensional (3D) T1-weighted imaging (with or without contrast enhancement), or axial and coronal T1- or T2-weighted imaging, slice thickness $\leq$ 3.0 mm (if available, with fat saturation))). The indication for surgery was either the presence of DON as a sight-threatening condition (following EUGOGO classification [3]) with no or only insufficient response to high-dose steroid therapy, or surgery was performed due to severe functional complaints or cosmetic reasons in cases with moderate-to-severe disease severity (following EUGOGO classification). In detail, in cases with moderate-to-severe disease severity and active (following clinical activity score (CAS) [3]) disease, surgery was performed in individual exceptional cases with severe functional complaints persisting despite high-dose steroids (such as severe lid retraction > 2 mm or exophthalmos > 3 mm with corneal exposure and signs of developing corneal breakdown (cases with insufficient responsiveness to lubricants), or severe retrobulbar pain). In cases with moderate-to-severe disease severity and inactive (following CAS) disease, surgery was performed for cosmetic and functional reasons (e.g., to correct exophthalmos and lid retraction, to decrease intraocular tension, to improve preexisting strabismus, etc.) for rehabilitation purposes. All cases were discussed at the surgical board of our interdisciplinary center for orbital diseases.

According to the literature, DON was then considered to be present in our cohort if at least one of the following criteria was fulfilled in GO patients (for which no other underlying cause was apparent either): impaired color vision, reduced visual acuity, optic disc swelling/atrophy, visual field defects or relative afferent pupillary defect [1,15].

Pterional orbital decompression was performed in a standardized fashion as previously described [16]. Prior to the volumetric analysis with a focus on the influence of orbital muscle volume, which is currently being conducted for the first time, data on the postoperative outcome of some of the cohort had already been published [16].

The primary endpoint of the current study was whether there were differences in the preoperative status and the postoperative development of visual acuity and exophthalmos depending on the orbital muscle volume ("not-increased" or "increased" according to the described subtypes [11]). The secondary endpoints were to determine the distribution of the muscular subtypes within the whole group and within the subgroups of cases with and without DON, as well as to assess other factors having potential influence, such as clinical activity or disease severity.

The retrospective evaluation was based on medical records and radiological imaging. Every surgery was performed in the Department of Neurosurgery as a tertiary referral center. Informed consent for the surgical procedure was obtained from all patients or their legal representative. The retrospective analysis was approved by the independent ethics committee of our medical center (reference no. 351/19) and has been reported according to institutional guidelines.

2.2. Volumetric Analysis

Volumetric analysis of intraorbital (extraocular) muscle volume and total orbital volume was performed in a semi-automatic technique with manual adjustment by two independent investigators using the SmartBrush device on the Brainlab Origin Server 3.2 (Brainlab AG, Munich, Germany), and the corresponding mean values were used for further analysis. Group classification was based on the calculated fraction of muscle volume relative to total orbital volume. The not-increased orbital muscle volume group (not-increased group) had a muscle fraction ≤ 0.19, and the increased orbital muscle volume group (increased group) had a muscle fraction ≥ 0.21 (following Regensburg et al., 2011 [11]). The volumetry technique using the Brainlab system is illustrated in Figure 1. MRI scans with examples of a not-increased and an increased orbital muscle volume are presented in Figure 2.

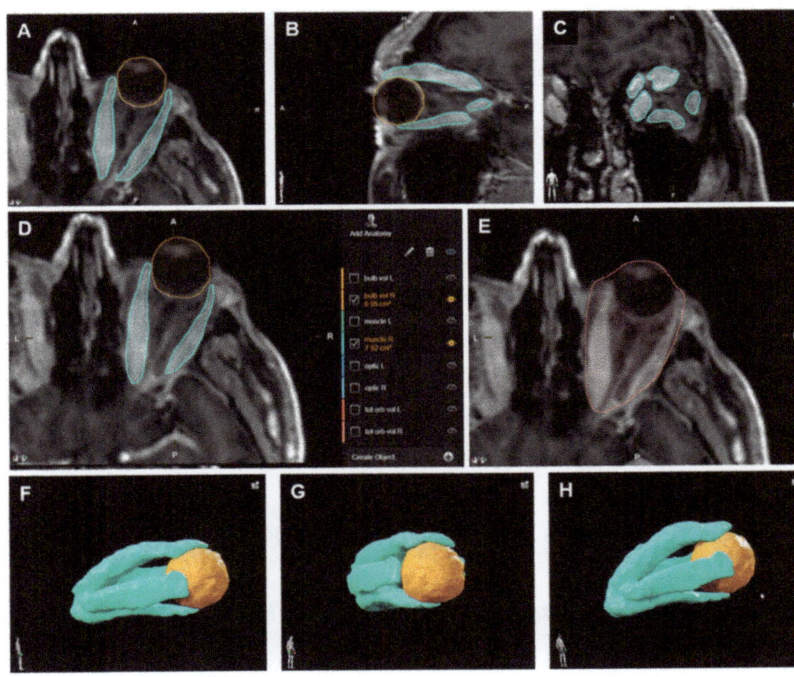

Figure 1. Illustration of the semi-automatic volumetry technique. (**A–C**): Volumetric analysis of the orbital content (right orbit) in a patient affected by Graves' orbitopathy (GO) (not-increased orbital muscle volume) prior to decompressive surgery, with orbital extraocular muscles marked in blue and eye bulb in orange in an axial (**A**), sagittal (**B**) and coronal (**C**) view. (**D**): Absolute volumetric values (in cm^3) provided by the system after three-dimensional (3D) analysis of the structures of interest (for example eye bulb with 6.95 cm^3 and muscle with 7.92 cm^3). (**E**): Volumetric analysis of the whole orbital content (marked in red in an axial view) for subsequent calculation of the orbital muscle fraction in relation to the total orbital volume. (**G,H**): 3D view of the volume-analyzed structures (orbital extraocular muscles marked in blue and eye bulb in orange), shown from different perspectives: lateral right (**F**), anterior oblique right (**G**), and posterior oblique right (**H**).

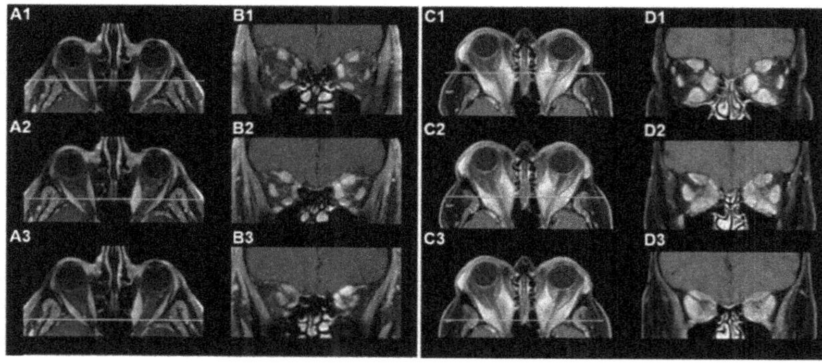

Figure 2. Magnetic resonance imaging (MRI) demonstrating not-increased or increased orbital muscle volume. (**A**,**B**): T1-weighted, contrast-enhanced MRI scans of the orbits of a bilaterally affected GO-patient with not-increased orbital muscle volume prior to decompressive surgery; axial images (**A1**–**A3**) with the orange line indicating the position of each corresponding coronal (**B1**–**B3**) image retrobulbar (**A1**,**B1**), at the orbital apex (**A3**,**B3**) and in between (**A2**,**B2**). (**C**,**D**): T1-weighted, contrast-enhanced MRI scans of the orbits of a bilaterally affected GO-patient with increased orbital muscle volume prior to decompressive surgery; axial images (**C1**–**C3**) with the orange line indicating the position of each corresponding coronal (**D1**–**D3**) image retrobulbar (**C1**,**D1**), at the orbital apex (**C3**,**D3**) and in between (**C2**,**D2**).

2.3. Ophthalmological Outcome

Preoperative ophthalmological data were obtained within the four weeks prior to surgery, and in acute sight-threatening cases, usually within the week prior to surgery. The measurement of visual acuity documented in the medical records in decimal notation was converted to the mean angle of resolution (logMAR) for analysis. It is important to note that better visual acuity correlates with lower logMAR values, and one line read on a standard vision chart corresponds to a difference of 0.1 logMAR. Exophthalmos assessments were performed by Hertel exophthalmometry. Within six months after surgery, the best documented values for visual acuity and exophthalmos were taken as postoperative values for analysis.

2.4. Statistical Analysis

Methods of descriptive statistics were used. Categorical data are presented as absolute and relative frequencies (in %). For numerical data, mean values with the minimum/maximum and the standard deviation (SD), standard error of mean (SEM) and confidence interval (CI) were calculated (as appropriate). Interrater reliability in volumetric analysis was validated using a Bland–Altman plot. Statistical differences were evaluated using a Mann–Whitney test as an unpaired nonparametric test. The level of significance was set to $p < 0.05$. Statistical analysis was performed using GraphPad Prism software version 9.1.1 for Mac (GraphPad Software, San Diego, CA, USA).

3. Results

3.1. Volumetry

The analysis of orbital muscle volume in relation to total orbital volume was performed in 91 orbits in 54 GO patients. Two groups could be classified in accordance with a "not-increased" orbital muscle fraction ($\leq$0.19) or an "increased" orbital muscle fraction ($\geq$0.21). Due to having an orbital muscle volume fraction between 0.19 and 0.21, eight orbits in seven patients were excluded. A total of 83 orbits in 47 patients were included in the study with an equal distribution among the groups (not-increased group: 42 orbits; increased group: 41 orbits). The mean orbital muscle volume fraction in the increased group was 29.63% (range 21.94–36.81%, SD 4.00, SEM 0.62, CI 28.36–30.89), which was significantly higher than in the not-increased group with a mean orbital muscle volume fraction of 15.60% (range 8.94–18.90%, SD 2.16, SEM 0.33, CI 14.93–16.28), $p < 0.0001$ (see Figure 3). The number of cases that were affected by DON (sight-threatening condition following EUGOGO classification) did not differ significantly between the two groups (increased group: 80.49% with DON, not-increased group: 71.43% with DON). The gender ratio was unequal in the not-increased group, with significantly more women than men affected (29:13, $p = 0.014$), whereas there was an equal distribution in the increased group (23:18). There were no significant differences between the groups regarding age, affected side, or clinical activity score (CAS). Detailed information is listed in Table 1.

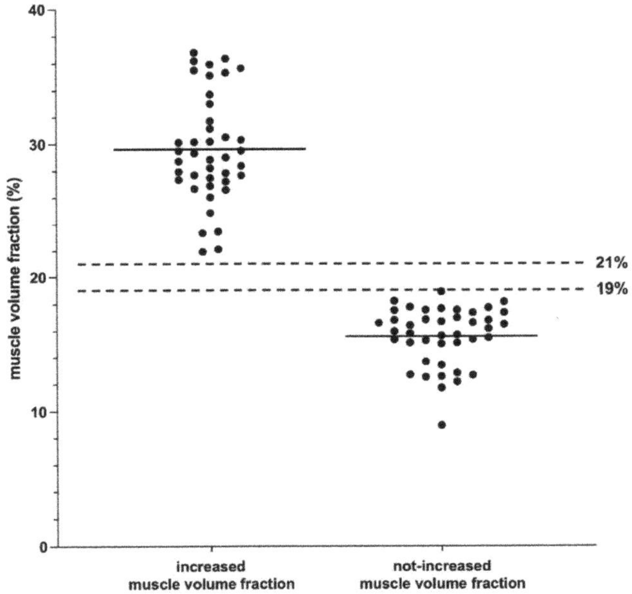

Figure 3. Volumetric analysis of orbital muscle volume fraction prior to orbital decompression. y axis: Orbital muscle volume fraction in relation to total orbital volume in %, x axis: group of orbits with an increased orbital muscle volume fraction (left, fraction $\geq$ 21%) and group of orbits with a not-increased orbital muscle volume fraction (right, fraction $\leq$ 19%), black dots: muscle volume fraction per orbit, horizontal black line: mean of values, dotted black lines: cut-off values (21% for increased muscle volume fraction and 19% for not-increased muscle volume fraction).

3.2. Outcome—Visual Acuity

Pre- and postoperative visual acuity measurements in accordance with the inclusion criteria were available for 35 of the 42 orbits in the not-increased group, and for 40 of the 41 orbits in the increased group. The mean preoperative visual acuity in the increased vs. the not-increased group was significantly reduced by 2.5 lines (logMAR 0.53 vs. 0.28, $p = 0.0058$). The mean postoperative visual acuity in the increased vs. the not-increased group was reduced by 2.1 lines, but without statistically significant difference. Within the increased group, the mean postoperative visual acuity was significantly improved by 1.7 lines compared with preoperative values (logMAR 0.36 vs. 0.53, $p = 0.0138$). Within the not-increased group, the mean postoperative visual acuity was improved by 1.3 lines compared with preoperative values, but without statistically significant difference. For a schematic illustration, please refer to Figures 4 and 5. A potential influence of the severity of proptosis on preoperative visual acuity was also analyzed in the cohort, but no relevant effect on visual acuity related to the severity of proptosis could be observed.

Table 1. Classification and demographic data of the "not-increased orbital muscle volume" and the "increased orbital muscle volume" group.

		Not-Increased Muscle Volume	Increased Muscle Volume
number of orbits		$n = 42$	$n = 41$
mean muscle volume (in relation to total orbital volume)		15.60% (SD 2.16, SEM 0.33, CI 14.93–16.28)	29.63% (SD 4.00, SEM 0.62, CI 28.36–30.89)
mean age (years)		55.33 (SD 11.08)	58.90 (SD 10.46)
gender	female	29 (69.05%)	23 (56.10%)
	male	13 (30.95%)	18 (43.90%)
side	left	22 (52.38%)	20 (48.78%)
	right	20 (47.62%)	21 (51.22%)
indication for surgery	DON	30 (71.43%)	33 (80.49%)
	cosmetic/functional	12 (28.57%)	8 (19.51%)
EUGOGO classification	moderate-to-severe	12 (28.57%)	8 (19.51%)
	sight-threatening	30 (71.43%)	33 (80.49%)
CAS	active	35 (83.33%)	38 (92.68%)
	inactive	7 (16.67%)	3 (7.32%)
diplopia	preoperatively	28 (66.67%)	29 (70.73%)
	postoperatively	26 (61.90%) (new after surgery: 2)	19 (46.34%) (new after surgery: 0)

SD, standard deviation; SEM, standard error of mean; CI, confidence interval; DON, dysthyroid optic neuropathy; EUGOGO, European group on Graves' orbitopathy; CAS, clinical activity score; postoperatively, within the six months after surgery.

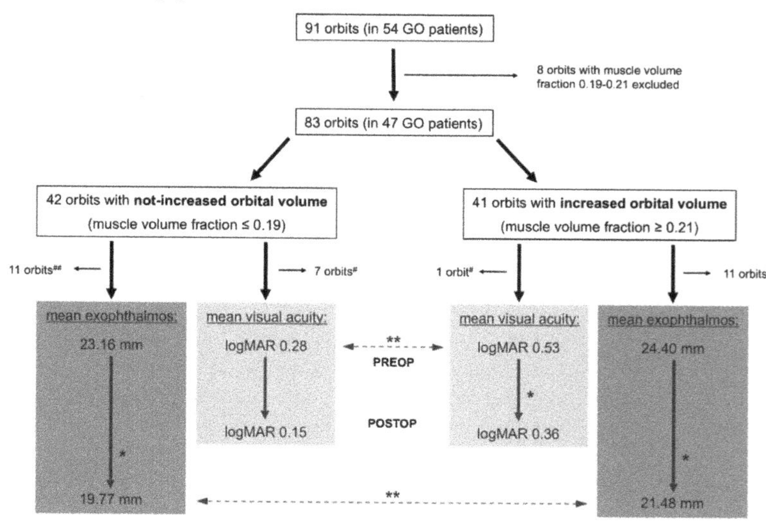

Figure 4. Inclusion criteria, group classification, and ophthalmological outcome. GO, Graves' orbitopathy; #, exclusion due to unavailability of postoperative assessment of visual acuity; ##, exclusion due to unavailability of postoperative assessment of exophthalmos; PREOP, preoperative assessments; POSTOP, postoperative assessments; logMAR, logarithm of the mean angle of resolution; *, statistically significant difference in intra-group analysis (comparison between pre-versus (vs.) postoperative results within the groups); **, statistically significant difference in inter-group analysis (comparison between preoperative or postoperative results between the two different groups).

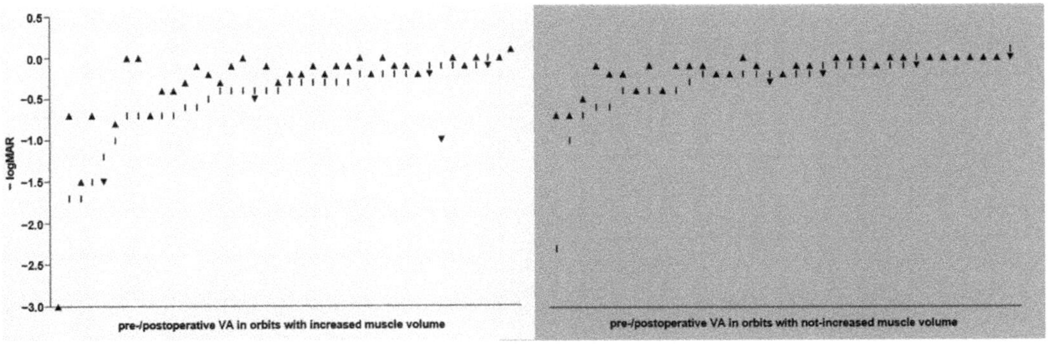

Figure 5. Development of visual acuity after orbital decompression. −logMAR, negative logarithm of the mean angle of resolution; VA, visual acuity; y axis, VA values per orbit before and after orbital decompression (given as negative logMAR, which means that the more negative the value, the worse the visual acuity); x axis, pre-/postoperative VA values of each orbit in the group with increased orbital muscle volume (left, light gray background) and in the group with not-increased orbital muscle volume (right, medium gray background); vertical line, preoperative VA value per orbit; black triangle (sharp angle upwards), postoperatively improved (or stable) VA value per orbit; black triangle (sharp angle downwards), postoperatively deteriorated VA value per orbit.

A total of 28 orbits in the not-increased group and 32 orbits in the increased group were affected by DON. Within the DON cohort, the mean preoperative visual acuity in the increased vs. the not-increased group was significantly reduced by 2.9 lines (mean logMAR 0.63 vs. 0.34, $p = 0.0035$). The mean post- vs. preoperative visual acuity was significantly improved in both the not-increased and the increased group affected by DON (not-increased group: logMAR 0.18 vs. 0.34 (1.6 lines, $p = 0.0447$), increased group: logMAR 0.43 vs. 0.63 (2.0 lines, $p = 0.0142$)). In the DON cohort, the mean postoperative visual acuity in the increased vs. the not-increased group was reduced by 2.5 lines without a statistically significant difference. Further information is given in Table 2.

3.3. *Outcome—Exophthalmos*

Pre- and postoperative exophthalmos measurements in accordance with the inclusion criteria were available for 31 of the 42 orbits in the not-increased group, and for 30 of the 41 orbits in the increased group. The mean postoperative exophthalmos of the not-increased vs. the increased group was significantly lower (a difference of 1.71 mm, $p = 0.0072$), whereas there was no statistically significant difference between the corresponding preoperative values (a difference of 1.24 mm). Both groups showed a significant reduction in their mean post- vs. preoperative exophthalmos (not-increased group: a difference of 3.39 mm ($p < 0.0001$); increased group: a difference of 2.92 mm ($p = 0.0082$)). For a schematic illustration, please refer to Figures 4 and 6.

The 24 orbits in each of the not-increased and the increased group affected by DON also presented a significant reduction in the mean exophthalmos postoperatively compared with the corresponding preoperative values (not-increased group: a difference of 3.55 mm ($p < 0.0001$); increased group: a difference of 3.13 mm ($p = 0.0309$)). In these orbits affected by DON, there were no significant differences between the not-increased and increased group regarding their mean pre- and postoperative exophthalmos values. Further information is given in Table 3.

Table 2. Intra- and inter-group statistical analysis of pre- and postoperative visual acuity (analysis within and between "not-increased" and "increased" orbital muscle volume).

		Not-Increased Muscle Volume	Increased Muscle Volume	Inter-Group Comparison: Change in VA (*p*-Value)
total group (DON + cosm/func)	no. of orbits	*n* = 35	*n* = 40	
	mean VA pre	0.28 logMAR {0.53 decimal} (SD 0.42, CI 0.14–0.43)	0.53 logMAR {0.30 decimal} (SD 0.59, CI 0.34–0.72)	***2.5 lines (p = 0.0058)***
	mean VA post	0.15 logMAR {0.72 decimal} (SD 0.19, CI 0.08–0.21)	0.36 logMAR {0.44 decimal} (SD 0.57, CI 0.18–0.54)	2.1 lines ($p = 0.0585$)
	change in VA post vs. pre (*p*-value)	1.3 lines ($p = 0.0761$)	***1.7 lines (p = 0.0138)***	
DON group	no. of orbits	*n* = 28	*n* = 32	
	mean VA pre	0.34 logMAR {0.46 decimal} (SD 0.45, CI 0.17–0.52)	0.63 logMAR {0.24 decimal} (SD 0.62, CI 0.41–0.85)	***2.9 lines (p = 0.0035)***
	mean VA post	0.18 logMAR {0.66 decimal} (SD 0.20, CI 0.10–0.25)	0.43 logMAR {0.38 decimal} (SD 0.62, CI 0.21–0.65)	2.5 lines ($p = 0.0578$)
	change in VA post vs. pre (*p*-value)	***1.6 lines (p = 0.0447)***	***2.0 lines (p = 0.0142)***	
cosm/func group	no. of orbits	*n* = 7	*n* = 8	
	mean VA pre	0.04 logMAR {0.92 decimal} (SD 0.10, CI −0.05–0.13)	0.11 logMAR {0.78 decimal} (SD 0.08, CI 0.04–0.18)	0.7 lines ($p = 0.1834$)
	mean VA post	0.04 logMAR {0.92 decimal} (SD 0.08, CI −0.03–0.12)	0.08 logMAR {0.84 decimal} (SD 0.07, CI 0.02–0.13)	0.4 lines ($p = 0.3930$)
	change in VA post vs. pre (*p*-value)	0.0 lines ($p > 0.9999$)	0.3 lines ($p = 0.5097$)	

VA, visual acuity; total group (DON + cosm/func), all cases with available pre- and postoperative assessment of VA in which dysthyroid optic neuropathy (DON) was the indication for surgery or which were operated due to cosmetic/functional reasons; DON group, all cases of "total group" in which DON was the indication for surgery; cosm/func group, all cases of "total group" which were operated due to cosmetic/functional reasons; no., number; vs., versus; change in VA post vs. pre, difference in post- vs. preoperative VA; logMAR, logarithm of the mean angle of resolution; pre, preoperatively; post, postoperatively (within six months after surgery); SD, standard deviation; CI, confidence interval; bold italic font, significant difference.

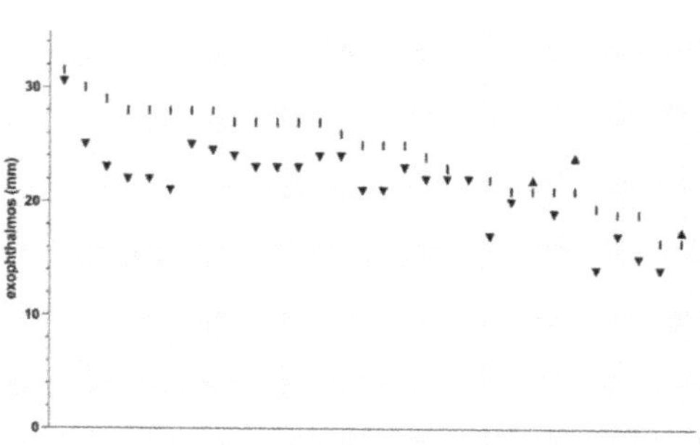

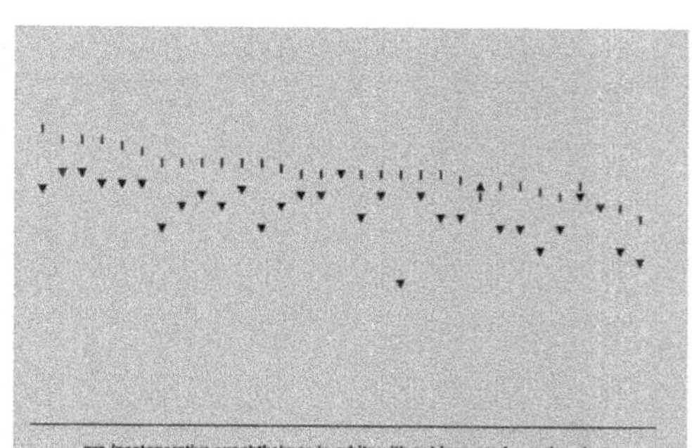

Figure 6. Development of exophthalmos after orbital decompression. y axis = exophthalmos values

(in mm) per orbit before and after orbital decompression, x axis = pre-/postoperative exophthalmos values of each orbit in the group with increased orbital muscle volume (left, light gray background) and in the group with not-increased orbital muscle volume (right, medium gray background), vertical line = preoperative exophthalmos value per orbit, black triangle (sharp angle downwards) = postoperative decrease in exophthalmos (or stable exophthalmos) per orbit, black triangle (sharp angle upwards) = postoperative increase in exophthalmos per orbit.

Table 3. Intra- and inter-group statistical analysis of pre- and postoperative exophthalmos (analysis within and between "not-increased" and "increased" orbital muscle volume).

		Not-Increased Muscle Volume	Increased Muscle Volume	Inter-Group Comparison: Change in Exoph (*p*-Value)
total group (DON + cosm/func)	no. of orbits	$n = 31$	$n = 30$	
	mean exoph pre (in mm)	23.16 (SD 1.88, CI 22.47–23.85)	24.40 (SD 4.04, CI 22.89–25.91)	1.24 mm ($p = 0.1140$)
	mean exoph post (in mm)	19.77 (SD 2.46, CI 18.87–20.67)	21.48 (SD 3.59, CI 20.14–22.82)	***1.71 mm (p = 0.0072)***
	change in exoph post vs. pre (*p*-value)	***3.39 mm (p < 0.0001)***	***2.92 mm (p = 0.0082)***	
DON group	no. of orbits	$n = 24$	$n = 24$	
	mean exoph pre (in mm)	23.38 (SD 1.81, CI 22.61–24.14)	24.17 (SD 4.39, CI 22.31–26.02)	0.79 mm ($p = 0.4564$)
	mean exoph post (in mm)	19.83 (SD 2.64, CI 18.72–20.95)	21.04 (SD 3.86, CI 19.41–22.67)	1.21 mm ($p = 0.1166$)
	change in exoph post vs. pre (*p*-value)	***3.55 mm (p < 0.0001)***	***3.13 mm (p = 0.0309)***	
cosm/func group	no. of orbits	$n = 7$	$n = 6$	
	mean exoph pre (in mm)	22.43 (SD 2.07, CI 20.51–24.34)	25.33 (SD 2.16, CI 23.07–27.60)	***2.90 mm (p = 0.0408)***
	mean exoph post (in mm)	19.57 (SD 1.81, CI 17.90–21.25)	23.25 (SD 1.08, CI 22.11–24.39)	***3.68 mm (p = 0.0058)***
	change in exoph post vs. pre (*p*-value)	***2.86 mm (p = 0.0262)***	2.08 mm ($p = 0.0909$)	

exoph, exophthalmos; total group (DON + cosm/func), all cases with available pre- and postoperative assessment of exophthalmos in which dysthyroid optic neuropathy (DON) was the indication for surgery or which were operated due to cosmetic/functional reasons; DON group, all cases of "total group" in which DON was the indication for surgery; cosm/func group, all cases of "total group" which were operated due to cosmetic/functional reasons; no., number; pre, preoperatively; post, postoperatively (within the six months after surgery); vs., versus; change in exoph post vs. pre, difference in post- vs. preoperative exophthalmos; SD, standard deviation; CI, confidence interval; bold italic font, significant difference.

3.4. Other Outcome Parameters

Diplopia was postoperatively completely resolved in 34.48% of cases in the increased group and in 14.29% of cases in the not-increased group. This difference did not reach statistical significance. New diplopia occurred postoperatively in two cases in the not-increased group (7.14%). There was transient headache after five operations (6%), periorbital hyposensitivity after two operations (2.4%) and clinical hemorrhage not affecting VA without further need for surgical treatment after one operation (1.2%).

4. Discussion

The visualization and analysis of the enlarged (intra-)orbital tissue in GO is of high importance in the understanding of the disease, particularly in severe cases and for the potential prediction of DON. Although MRI is also used for imaging diagnostics today, CT continues to be frequently used as the preferred technique since both the bony orbit and intraorbital structures are excellently visualized [1,6,17]. Volumetric analysis of (intra-)orbital structures has been established on the basis of both CT and MR images with comparable results [8,14,18–20]. In addition to fully manual segmentation (planimetry), the semi-automatic technique (with the possibility of manual adjustment) has also proven successful for volumetric analysis [21,22].

In studies of the natural disease course in untreated GO patients, fat enlargement tends to be classified as a late phenomenon, and muscle enlargement is rather associated with more severe disease activity [7,10,12]. An increased muscle volume in relation to total orbital volume was defined as a muscle fraction of 0.21 or more [10,11]. Regensburg et al., described four subtypes of GO based on CT volumetric analysis of 95 untreated GO orbits (25% without fat/muscle increase, 5% with only fat increase, 61% with only muscle increase, 9% with both fat and muscle increase) [11]. Thus, 69.5% of these orbits showed muscle enlargement with a mean muscle fraction of 0.24–0.25 (relative to total orbital volume), whereas orbits with "not-increased muscle volume" had a mean muscle fraction of 0.16–0.17. In our analysis with analogous definitions of "not-increased volume" and "increased volume" there was a highly significant difference between the group with "increased" mean muscle volume (fraction 0.29–0.30) and "not-increased" mean muscle volume (fraction 0.15–0.16). For healthy control groups, data on orbital muscle fraction range from 0.12 to 0.15 [7,23].

To the best of our knowledge, the current study is the first to examine orbital muscle enlargement as a potential factor that influences functional outcome in a large cohort of GO patients who were all treated with the same surgical approach. The choice of the approach for orbital decompression is discussed controversially in the literature. Despite a number of systematic reviews and also prospective studies, no general recommendations on the most suitable surgical procedure have yet been established due to various surgical indications and the heterogeneity of applied techniques. In addition, surgeons of different specialties have varying preferences depending on their personal skills and experience [24,25].

In our study, there was a balanced distribution of orbits with increased muscle volume and those with not-increased muscle volume. Furthermore, the number of cases affected by DON was not significantly higher in the increased group compared with the not-increased group. One might have suspected that there would be a majority of cases with increased muscle volume in a cohort of severe cases requiring surgical treatment, as this seems to be associated with more severe disease [6,7,10,12,26]. However, our results support the hypothesis that the absence of an increased muscle volume does not imply a less severe course or that DON is necessarily associated with an increased muscle volume [12,13]. Nevertheless, muscle diameter and volume at the orbital apex is of particular interest in the context of DON ("apical crowding") [27–30]. Several studies have shown a significant difference in apical crowding between orbits with and without DON [5,6,12]. However, an analysis of the functional outcome after surgical therapy depending on these morphological differences is still missing.

In our analysis, the mean preoperative visual acuity was significantly lower in the increased vs. the not-increased group, also in the subgroup of DON-affected orbits. Although the number of orbits affected by DON itself was not significantly different between the increased and the not-increased group, an increased muscle volume nevertheless resulted in significantly worse preoperative visual acuity. Interestingly, however, decompression led to a significant improvement in visual acuity in the increased group, so after surgery there was no longer a significant difference between the mean visual acuity of the increased vs. the not-increased group. Presumably, the lower extent of postoperative improvement of visual acuity in the group with a not-increased orbital muscle volume fraction is related to the fact that preoperative visual acuity was less reduced in this group. In the subgroup of DON-affected orbits, surgical therapy resulted in a significant improvement in visual acuity in both the not-increased and the increased group. Again, the improvement in the increased group was such that the preoperative significant difference in mean visual acuity between the groups was eliminated postoperatively.

Thus, there was a strong association between increased orbital muscle volume and lower visual acuity, and remarkably, there was an excellent response to surgical decompression, even in cases with increased orbital muscle volume. As a predictive factor for DON-affected orbits, increased muscle volume was found to be associated with significantly lower visual acuity, and in view of the good response to surgical therapy, decompression should be considered at an early stage, especially in this group.

Along with varying surgical approaches, several individual factors are discussed as possible reasons for the different outcomes of orbital decompression in GO [31]. Morphological factors seem to be particularly important here, such as the shape and angle of the bony orbit and the anatomy of the venous and lymphatic vessels [31–33]. Few studies have investigated factors influencing the visual outcome after orbital decompression in GO, such as duration of the disease, preoperative visual acuity, additional orbital fat reduction, and preoperative degree of exophthalmos [16,34–36]. CT-based volumetric analyses showed a positive correlation between exophthalmos reduction and decompression volume [37,38]. A recent study designed a phantom model to measure the pressure relief at the orbital apex depending on the localization and extent of bony decompression comparing different surgical techniques for orbital decompression. This concept seems very interesting and should be further pursued and expanded, as it provides an additional factor that has not yet been considered in the choice of surgical approaches [39].

In our study, the mean preoperative exophthalmos in the increased group (total group as well as subgroup affected by DON) was not significantly higher than in the not-increased group, and decompression resulted in a significant exophthalmos decrease in both groups. This decrease was more pronounced in the not-increased group, which is why the persistent mean postoperative exophthalmos was significantly higher in orbits with increased than with not-increased orbital muscle volume. Hence, no clear association could be established between increased orbital muscle volume and a higher degree of preoperative exophthalmos, but exophthalmos was significantly more reduced after decompression when there was not-increased orbital muscle volume compared with increased orbital muscle volume.

The strengths of our study include the large number of cases, the design having a clear division into two groups based on the determined muscle volume with subsequent standardized surgical treatment, and the objective outcome parameters. Limitations arise in the retrospective design, and the need to exclude cases in the absence of appropriate preoperative imaging or missing outcome data. Furthermore, in a retrospective analysis of over 20 years, surgeon-dependent differences may naturally occur despite the standardized surgical procedure, which could have an influence on the decompression volume. This needs to be further investigated in subsequent studies.

5. Conclusions

The development of dysthyroid optic neuropathy in Graves' orbitopathy is not necessarily related to orbital muscle enlargement. Increased orbital muscle volume is associated with a significant reduction in preoperative visual acuity, but not with increased preoperative exophthalmos. An excellent improvement in visual acuity can be achieved by pterional orbital decompression, even with increased orbital muscle volume. Therefore, this factor should be addressed during treatment decisions, and early surgical therapy should be considered particularly in DON-affected cases with orbital muscle enlargement.

Author Contributions: Conceptualization, C.S. and J.G.; methodology, C.S. and J.G.; software, C.S. and S.K.; validation, C.S., W.M. and J.G.; formal analysis, C.S. and W.M.; investigation, C.S., W.M. and S.K.; resources, C.S. and J.G.; data curation, C.S. and S.K.; writing—original draft preparation, C.S.; writing—review and editing, C.S., J.G., J.B. and W.A.L.; visualization, C.S. and S.K.; supervision, J.G., J.B. and W.A.L.; project administration, J.G.; funding acquisition, C.S. All authors have read and agreed to the published version of the manuscript.

Funding: This research received no external funding.

Institutional Review Board Statement: The study was conducted in accordance with the Declaration of Helsinki, and approved by the Institutional Independent Ethics Committee of the Albert-Ludwigs-University Freiburg, Germany (reference no. 351/19, 27 August 2019).

Informed Consent Statement: Informed consent for the surgical procedure was obtained from all patients or their legal representative.

Data Availability Statement: The data used to support the findings of this study are available from the corresponding author upon request.

Acknowledgments: The article processing charge was funded by the Ministry for Science, Research and Arts, Baden-Wuerttemberg, and the University of Freiburg in the funding programme Open Access Publishing.

Conflicts of Interest: The authors declare no conflict of interest.

References

1. McKeag, D.; Lane, C.; Lazarus, J.H.; Baldeschi, L.; Boboridis, K.; Dickinson, A.J.; Hullo, A.I.; Kahaly, G.; Krassas, G.; Marcocci, C.; et al. Clinical Features of Dysthyroid Optic Neuropathy: A European Group on Graves' Orbitopathy (EUGOGO) Survey. *Br. J. Ophthalmol.* **2007**, *91*, 455–458. [CrossRef] [PubMed]
2. Bahn, R.S. Graves' Ophthalmopathy. *N. Engl. J. Med.* **2010**, *362*, 726–738. [CrossRef]
3. Bartalena, L.; Kahaly, G.J.; Baldeschi, L.; Dayan, C.M.; Eckstein, A.; Marcocci, C.; Marinò, M.; Vaidya, B.; Wiersinga, W.M.; EUGOGO. The 2021 European Group on Graves' Orbitopathy (EUGOGO) Clinical Practice Guidelines for the Medical Management of Graves' Orbitopathy. *Eur. J. Endocrinol.* **2021**, *185*, G43–G67. [CrossRef]
4. Bartalena, L.; Baldeschi, L.; Dickinson, A.; Eckstein, A.; Kendall-Taylor, P.; Marcocci, C.; Mourits, M.; Perros, P.; Boboridis, K.; Boschi, A.; et al. Consensus Statement of the European Group on Graves' Orbitopathy (EUGOGO) on Management of GO. *Eur. J. Endocrinol.* **2008**, *158*, 273–285. [CrossRef]
5. Chan, L.-L.; Tan, H.-E.; Fook-Chong, S.; Teo, T.-H.; Lim, L.-H.; Seah, L.-L. Graves Ophthalmopathy: The Bony Orbit in Optic Neuropathy, Its Apical Angular Capacity, and Impact on Prediction of Risk. *Am. J. Neuroradiol.* **2009**, *30*, 597–602. [CrossRef]
6. Gonçalves, A.C.P.; Silva, L.N.; Gebrim, E.M.M.S.; Monteiro, M.L.R. Quantification of Orbital Apex Crowding for Screening of Dysthyroid Optic Neuropathy Using Multidetector CT. *Am. J. Neuroradiol.* **2012**, *33*, 1602–1607. [CrossRef]
7. Wiersinga, W.M.; Regensburg, N.I.; Mourits, M.P. Differential Involvement of Orbital Fat and Extraocular Muscles in Graves' Ophthalmopathy. *Eur. Thyroid. J.* **2013**, *2*, 14–21. [CrossRef] [PubMed]
8. Keene, K.R.; van Vught, L.; van de Velde, N.M.; Ciggaar, I.A.; Notting, I.C.; Genders, S.W.; Verschuuren, J.J.G.M.; Tannemaat, M.R.; Kan, H.E.; Beenakker, J.-W.M. The Feasibility of Quantitative MRI of Extra-Ocular Muscles in Myasthenia Gravis and Graves' Orbitopathy. *NMR Biomed.* **2021**, *34*, e4407. [CrossRef]
9. Nagy, E.V.; Toth, J.; Kaldi, I.; Damjanovich, J.; Mezosi, E.; Lenkey, A.; Toth, L.; Szabo, J.; Karanyi, Z.; Leovey, A. Graves' Ophthalmopathy: Eye Muscle Involvement in Patients with Diplopia. *Eur. J. Endocrinol.* **2000**, *142*, 591–597. [CrossRef]
10. Potgieser, P.W.; Wiersinga, W.M.; Regensburg, N.I.; Mourits, M.P. Some Studies on the Natural History of Graves' Orbitopathy: Increase in Orbital Fat Is a Rather Late Phenomenon. *Eur. J. Endocrinol.* **2015**, *173*, 149–153. [CrossRef] [PubMed]
11. Regensburg, N.I.; Wiersinga, W.M.; Berendschot, T.T.J.M.; Potgieser, P.; Mourits, M.P. Do Subtypes of Graves' Orbitopathy Exist? *Ophthalmology* **2011**, *118*, 191–196. [CrossRef]
12. Al-Bakri, M.; Rasmussen, A.K.; Thomsen, C.; Toft, P.B. Orbital Volumetry in Graves' Orbitopathy: Muscle and Fat Involvement in Relation to Dysthyroid Optic Neuropathy. *ISRN Ophthalmol.* **2014**, *2014*, 435276. [CrossRef] [PubMed]
13. Anderson, R.L.; Tweeten, J.P.; Patrinely, J.R.; Garland, P.E.; Thiese, S.M. Dysthyroid Optic Neuropathy without Extraocular Muscle Involvement. *Ophthalmic Surg.* **1989**, *20*, 568–574. [CrossRef]
14. Zou, M.; Wu, D.; Zhu, H.; Huang, X.; Zhao, X.; Zhao, J.; Fu, W.; Li, R.; Li, B.; Wan, P.; et al. Multiparametric Quantitative MRI for the Evaluation of Dysthyroid Optic Neuropathy. *Eur. Radiol.* **2021**, *32*, 1931–1938. [CrossRef]
15. Dayan, C.M.; Dayan, M.R. Dysthyroid Optic Neuropathy: A Clinical Diagnosis or a Definable Entity? *Br. J. Ophthalmol.* **2007**, *91*, 409–410. [CrossRef]
16. Kuechlin, S.; Steiert, C.; Naseri, Y.; Joachimsen, L.; Gruber, M.; Reich, M.; Boehringer, D.; Metzger, M.; Beck, J.; Scheiwe, C.; et al. Pterional Orbit Decompression in Grave Disease with Dysthyroid Optic Neuropathy. *World Neurosurg.* **2021**, *149*, e1007–e1016. [CrossRef]

17. Smith, T.J.; Hegedüs, L. Graves' Disease. *N. Engl. J. Med.* **2016**, *375*, 1552–1565. [CrossRef]
18. Willaert, R.; Degrieck, B.; Orhan, K.; Deferm, J.; Politis, C.; Shaheen, E.; Jacobs, R. Semi-Automatic Magnetic Resonance Imaging Based Orbital Fat Volumetry: Reliability and Correlation with Computed Tomography. *Int. J. Oral Maxillofac. Surg.* **2021**, *50*, 416–422. [CrossRef]
19. Regensburg, N.I.; Kok, P.H.B.; Zonneveld, F.W.; Baldeschi, L.; Saeed, P.; Wiersinga, W.M.; Mourits, M.P. A New and Validated CT-Based Method for the Calculation of Orbital Soft Tissue Volumes. *Investig. Ophthalmol. Vis. Sci.* **2008**, *49*, 1758–1762. [CrossRef]
20. Bijlsma, W.R.; Mourits, M.P. Radiologic Measurement of Extraocular Muscle Volumes in Patients with Graves' Orbitopathy: A Review and Guideline. *Orbit* **2006**, *25*, 83–91. [CrossRef]
21. Sentucq, C.; Schlund, M.; Bouet, B.; Garms, M.; Ferri, J.; Jacques, T.; Nicot, R. Overview of Tools for the Measurement of the Orbital Volume and Their Applications to Orbital Surgery. *J. Plast. Reconstr. Aesthet Surg.* **2021**, *74*, 581–591. [CrossRef]
22. Bontzos, G.; Mazonakis, M.; Papadaki, E.; Maris, T.G.; Blazaki, S.; Drakonaki, E.E.; Detorakis, E.T. Ex Vivo Orbital Volumetry Using Stereology and CT Imaging: A Comparison with Manual Planimetry. *Eur. Radiol.* **2019**, *29*, 1365–1374. [CrossRef]
23. Regensburg, N.I.; Wiersinga, W.M.; van Velthoven, M.E.J.; Berendschot, T.T.J.M.; Zonneveld, F.W.; Baldeschi, L.; Saeed, P.; Mourits, M.P. Age and Gender-Specific Reference Values of Orbital Fat and Muscle Volumes in Caucasians. *Br. J. Ophthalmol.* **2011**, *95*, 1660–1663. [CrossRef]
24. Leong, S.C.; Karkos, P.D.; Macewen, C.J.; White, P.S. A Systematic Review of Outcomes Following Surgical Decompression for Dysthyroid Orbitopathy. *Laryngoscope* **2009**, *119*, 1106–1115. [CrossRef]
25. European Group on Graves' Orbitopathy (EUGOGO); Mourits, M.P.; Bijl, H.; Altea, M.A.; Baldeschi, L.; Boboridis, K.; Currò, N.; Dickinson, A.J.; Eckstein, A.; Freidel, M.; et al. Outcome of Orbital Decompression for Disfiguring Proptosis in Patients with Graves' Orbitopathy Using Various Surgical Procedures. *Br. J. Ophthalmol.* **2009**, *93*, 1518–1523. [CrossRef]
26. Barrett, L.; Glatt, H.J.; Burde, R.M.; Gado, M.H. Optic Nerve Dysfunction in Thyroid Eye Disease: CT. *Radiology* **1988**, *167*, 503–507. [CrossRef]
27. Giaconi, J.A.; Kazim, M.; Rho, T.; Pfaff, C. CT Scan Evidence of Dysthyroid Optic Neuropathy. *Ophthalmic Plast. Reconstr. Surg.* **2002**, *18*, 177–182. [CrossRef]
28. Hallin, E.S.; Feldon, S.E. Graves' Ophthalmopathy: I. Simple CT Estimates of Extraocular Muscle Volume. *Br. J. Ophthalmol.* **1988**, *72*, 674–677. [CrossRef]
29. Nugent, R.A.; Belkin, R.I.; Neigel, J.M.; Rootman, J.; Robertson, W.D.; Spinelli, J.; Graeb, D.A. Graves Orbitopathy: Correlation of CT and Clinical Findings. *Radiology* **1990**, *177*, 675–682. [CrossRef]
30. Weis, E.; Heran, M.K.S.; Jhamb, A.; Chan, A.K.; Chiu, J.P.; Hurley, M.C.; Rootman, J. Quantitative Computed Tomographic Predictors of Compressive Optic Neuropathy in Patients with Thyroid Orbitopathy: A Volumetric Analysis. *Ophthalmology* **2012**, *119*, 2174–2178. [CrossRef]
31. Stan, M.N.; Bahn, R.S. Risk Factors for Development or Deterioration of Graves' Ophthalmopathy. *Thyroid* **2010**, *20*, 777–783. [CrossRef] [PubMed]
32. Baujat, B.; Krastinova, D.; Bach, C.A.; Coquille, F.; Chabolle, F. Orbital Morphology in Exophthalmos and Exorbitism. *Plast. Reconstr. Surg.* **2006**, *117*, 542–550; discussion 551–552. [CrossRef]
33. Wiersinga, W.M.; Smit, T.; van der Gaag, R.; Mourits, M.; Koornneef, L. Clinical Presentation of Graves' Ophthalmopathy. *Ophthalmic Res.* **1989**, *21*, 73–82. [CrossRef]
34. Nanda, T.; Dunbar, K.E.; Campbell, A.A.; Bathras, R.M.; Kazim, M. Greater Proptosis Is Not Associated With Improved Compressive Optic Neuropathy in Thyroid Eye Disease. *Ophthalmic Plast. Reconstr. Surg.* **2018**, *34*, S72–S74. [CrossRef] [PubMed]
35. Wang, Y.; Li, Y.Y.; Yang, N.; Ma, R.; Xiao, L.H. Therapeutic outcomes and influence factors of maximal orbital decompression in the treatment of severe dysthyroid optic neuropathy. *Zhonghua Yan Ke Za Zhi* **2017**, *53*, 416–423. [CrossRef]
36. Wu, C.Y.; Stacey, A.W.; Kahana, A. Simultaneous Versus Staged Balanced Decompression for Thyroid-Related Compressive Optic Neuropathy. *Ophthalmic Plast. Reconstr. Surg.* **2016**, *32*, 462–467. [CrossRef]
37. Kim, K.W.; Byun, J.S.; Lee, J.K. Surgical Effects of Various Orbital Decompression Methods in Thyroid-Associated Orbitopathy: Computed Tomography-Based Comparative Analysis. *J. Craniomaxillofac. Surg.* **2014**, *42*, 1286–1291. [CrossRef] [PubMed]
38. Pereira, T.d.S.; Leite, C.d.A.; Kuniyoshi, C.H.; Gebrim, E.M.M.S.; Monteiro, M.L.R.; Pieroni Gonçalves, A.C. A Randomized Comparative Study of Inferomedial vs. Balanced Orbital Decompression. Analysis of Changes in Orbital Volume, Eyelid Parameters, and Eyeball Position. *Eye* **2021**, *36*, 547–554. [CrossRef]
39. Yo, K.; Nishimura, K.; Takahashi, Y.; Yokota, H.; Hatayama, N.; Hoshino, T.; Naito, M.; Ogawa, T.; Fujimoto, Y. Comparison of the Decompressive Effect of Different Surgical Procedures for Dysthyroid Optic Neuropathy Using 3D Printed Models. *Graefes Arch. Clin. Exp. Ophthalmol.* **2022**, *online ahead of print*. [CrossRef]

Journal of
Personalized Medicine

MDPI

Article

Change of Optical Coherence Tomography Morphology and Associated Structural Outcome in Diabetic Macular Edema after Ranibizumab Treatment

Nan-Ni Chen [1], Chien-Hsiung Lai [1,2,3,*], Chai-Yi Lee [4], Chien-Neng Kuo [5], Ching-Lung Chen [1,6], Jou-Chen Huang [7,8], Pei-Chen Wu [9], Pei-Lun Wu [1] and Chau-Yin Chen [1]

1 Department of Ophthalmology, Chang Gung Memorial Hospital, Chiayi 61363, Taiwan; nani4chen@gmail.com (N.-N.C.); jackeichen@gmail.com (C.-L.C.); peilun@cgmh.org.tw (P.-L.W.); ccy423@gmail.com (C.-Y.C.)
2 Department of Nursing, Chang Gung University of Science and Technology, Chiayi 61363, Taiwan
3 School of Traditional Chinese Medicine, College of Medicine, Chang Gung University, Taoyuan 33302, Taiwan
4 Department of Ophthalmology, Show Chwan Hospital, Changhua 50861, Taiwan; ao6u.3msn@hotmail.com
5 Department of Ophthalmology, Changhua Christian Hospital, Changhua 50861, Taiwan; terrry339@yahoo.com.tw
6 Department of Optometry, Chung Hwa University of Medical Technology, Tainan 71703, Taiwan
7 Department of Ophthalmology, Taipei Medical University Hospital, Taipei 110301, Taiwan; roro691213@gmail.com
8 Department of Ophthalmology, School of Medicine, College of Medicine, Taipei Medical University, Taipei 110301, Taiwan
9 Department of Ophthalmology, Landseed International Hospital, Taoyuan 32449, Taiwan; heyshowjim77@gmail.com
* Correspondence: oph4557@gmail.com; Tel.: +886-5362-1000 (ext. 3716)

Citation: Chen, N.-N.; Lai, C.-H.; Lee, C.-Y.; Kuo, C.-N.; Chen, C.-L.; Huang, J.-C.; Wu, P.-C.; Wu, P.-L.; Chen, C.-Y. Change of Optical Coherence Tomography Morphology and Associated Structural Outcome in Diabetic Macular Edema after Ranibizumab Treatment. *J. Pers. Med.* **2022**, *12*, 611. https://doi.org/10.3390/jpm12040611

Academic Editor: Chieh-Chih Tsai

Received: 26 March 2022
Accepted: 9 April 2022
Published: 11 April 2022

Abstract: (1) Background: To investigate the correlation between therapeutic outcome and morphologic changes for diabetic macular edema (DME) after intravitreal injection of ranibizumab (IVIR). (2) Methods: This retrospective study included 228 eyes received IVIR for DME. Each participant was traced for two years after the initial IVIR, while the data of ophthalmic examination, optical coherence tomography (OCT) image, and systemic diseases were collected. The study population was categorized into different subgroups according to the existence of OCT morphologic change and the initial OCT morphologic pattern, including diffuse retinal thickening (DRT), cystoid macular edema (CME), serous retinal detachment (SRD), and vitreomacular interface abnormalities (VMIAs). The primary outcomes were the baseline best-corrected visual acuity (BCVA) and central macular thickness (CMT) during a two-year study period. The distribution of OCT morphologic change and its relation to primary outcome were analyzed. (3) Results: Comparing the 42 eyes (18.4%) with OCT morphological changes to another 186 eyes (81.6%) without such alteration, the former showed a poorer baseline BCVA (0.84 ± 0.39 vs. 0.71 ± 0.36, $p = 0.035$), worse final BCVA (0.99 ± 0.44 vs. 0.67 ± 0.30, $p = 0.001$), and thicker final CMT (354.21 ± 89.02 vs. 305.33 ± 83.05, $p = 0.001$). Moreover, the VMIA developed in 14.9% of all DME patients presenting the most common morphologic change among DRT, CME, and SRD. Besides, the presence of stroke was independently correlated to the morphologic change (adjusted odds ratio [aOR]: 6.381, 95% confidence interval (CI): 1.112–36.623, $p = 0.038$). (4) Conclusions: The change of OCT morphology in DME patients receiving IVIR was correlated to worse structural and visual outcome while the formation of VMIA most commonly occurred after initial treatment.

Keywords: optical coherence tomography; diabetic macular edema; antivascular endothelial growth factors; vitreomacular interface abnormality

1. Introduction

Spectral domain optical coherence tomography (SD-OCT) has been rapidly developed and widely used for detection of alternation and quantification of retinal structure in various types of macular edema [1]. Several patterns of diabetic macular edema (DME) as well as OCT parameters have been proposed on SD-OCT, including diffuse retinal thickening (DRT), cystoid macular edema (CME), serous retinal detachment (SRD), and vitreomacular interface abnormalities (VMIAs) [1–8].

As intravitreous anti-vascular endothelial growth factors (anti-VEGF) injections have become the first-line therapy for DME [9,10], OCT patterns used in monitoring the effect of therapies for macular edema contribute to understanding the retinal anatomical response and structure damage of DME with distinctive aspects in each morphologic subtype [8]. Previous studies have reported that the effect of intravitreal anti-VEGF treatment is predictable among different patterns of DME [3,5–7]; however, seldom studies demonstrated the changes in OCT parameters over time after anti-VEGF therapy [11,12]. Although certain OCT morphology may be likely developed after anti-VEGF treatment for DME, the exact types need further validation.

Therefore, we aimed to investigate the association among different morphologic subtypes of DME and the response and changes in OCT morphological features over time after intravitreal ranibizumab (IVIR) injections.

2. Materials and Methods

2.1. Subject Selection

The retrospective cohort study was conducted and the medical records of 652 eyes were reviewed from 433 patients with DME receiving at least one IVIR with 0.5 mg of ranibizumab (0.05 mL) during February 2013 and June 2019 in Chang Gung Memorial Hospital, Chiayi branch in southern Taiwan. The inclusion criteria included (1) clinically significant macular edema (CSME) as defined by ETDRS; and (2) CMT of >300 μm as documented on OCT. The exclusion criteria were as follows: (1) any other ophthalmic disease apart from diabetic retinopathy and cataract; (2) previous intravitreal bevacizumab, intravitreal aflibercept, or intravitreal steroid injections; (3) previous macular focal laser therapy; (4) intraocular surgery or panretinal photocoagulation within 3 months of commencement of the study period; and (5) any missing data or poor OCT image quality at designated time points. After the selection process, a total of 228 eyes of 150 patients with OCT scans comprised the study population for evaluation of OCT morphologic characteristics.

2.2. Data Collection

At first visit, all patients underwent complete ophthalmic examination, including best-corrected Snellen visual acuity (BCVA), slit-lamp examination, intraocular pressure measurement, fundus photography, retina thickness measurement by OCT, and fluorescence angiography. BCVA values were converted to the logarithm of the minimum angle of resolution (logMAR) scale for statistical manipulation. Metabolic parameters were also assessed through review of the comprehensive medical records, including hypertension (HTN), coronary artery disease (CAD), stroke, chronic kidney disease (CKD), hyperlipidemia, thyroid disease, and cancer. After the IVIR, patients were examined monthly until 3 months, then 6, 12, and 24 months for collecting the OCT morphology parameters. The decision to perform reinjection was based on persistent or increased sub-retinal fluid/cystoid macular edema, or CMT > 300 μm on OCT at the monthly control visits. If any morphology change occurred, it would be recorded at all designated time points.

2.3. Morphology Measurement

One SD-OCT device (Optovue Inc., Fremont, CA, USA) with software version 6.2.2.73 was applied for the image collection and analysis. The eyes were classified into 4 subgroups based on the cross-sectional retinal morphologies. The anatomic outcome was defined as CMT reduction between the baseline and the control visit. The CMT was automatically calculated as the average retinal thickness within the central circle of a 500-μm radius. At the designated period of time of 2 years, all patients were divided into two groups according to the presence of morphology changes. Patients with any OCT morphology changes were categorized as group 1 (42 eyes), while those without changes of OCT pattern were regarded as group 2 (186 eyes). The OCT patterns were classified into four types: (1) the DRT pattern characterized by a generalized, heterogeneous, sponge-like swelling of the macula with mild hyporeflectivity compared with normal retina; (2) the CME pattern defined by intraretinal oval cystoid spaces of low reflectivity, typically separated by highly reflective septae; (3) the SRD pattern showing a hyporeflective, dome-shaped detachment between the retinal pigment epithelium (RPE) and the retina; and (4)VMIA including the presence of epiretinal membranes (ERM) or vitreomacular traction (VMT). Both ERM and VMT presented as a highly reflective band or membrane on the retinal surface. All patients were divided into four groups based on the OCT findings, as our previous work have mentioned [13]. DRT group included patients with only pure DRT, while CME group included patients with CME, but no subretinal fluid or vitreous macular interface abnormalities. If DRT and CME coexisted, then CME was considered the dominant pattern. SRD group comprised patients with SRD, but without VMIA. When DRT, CME, and SRD all presented, the eye was classified into SRD group. Regardless of pattern combinations, eyes with ERM or vitreomacular traction were classified into VMIA group.

2.4. Statistical Analysis

All analyses were computed by using the PASW Statistics 18 software (Version 18.0. SPSS Inc., Chicago, IL, USA). Descriptive analysis was used to show the baseline characteristics of the study group. Then, the basic properties as well as the changes of CMT between group 1 and group 2 were analyzed by using independent *t*-test and Chi-square test, respectively. Next, the rate of morphological changes among different OCT subgroups was analyzed by one-way analysis of variance with post-hoc exam of Dunnett T3. The bar chart was plotted to demonstrate the distribution of OCT morphology change among the different subgroups. Besides, the multivariate logistic regression analysis was utilized to yield adjusted odds ratio (aOR), corresponding 95% confidence interval (CI) for evaluating the correlation between OCT morphology alteration and demographic data as well as systemic parameters. A p value of less than 0.05 is regarded as statistical significance.

3. Results

3.1. Baseline Characteristics and Treatment Outcome

The baseline characteristics of the study population are summarized in Table 1. The mean age was 66.30 ± 9.47 years in the group 1 and 65.66 ± 9.01 in group 2 without significant difference ($p = 0.720$). The overall average dosages of injections for all patients were 4.31 ± 1.26 in the first year and 5.33 ± 1.90 in two years. Besides, no differences were noted between group 1 and group 2 regarding the baseline characteristics of gender, diabetes retinopathy stage (PDR/NPDR), panretinal photocoagulation, baseline CMT and average one-year and two-year injection dosage (all $p > 0.05$). However, the baseline BCVA, final BCVA, and CMT at second year of treatment were significantly different between two groups ($p = 0.035$, 0.001, and 0.001, respectively) in which group 1 was correlated to a poorer baseline and final BCVA (0.84 ± 0.39 and 0.99 ± 0.44, log MAR), and greater second year CMT (354.21 ± 89.02) (Table 1).

Table 1. Demographic and clinical properties of the patients.

Description	Group 1	Group 2	p-Value
No. of eyes	42	186	
Male/Female	19/14	76/41	0.437
Age, years	66.30 ± 9.47	65.66 ± 9.01	0.720
Baseline BCVA (logMAR)	0.84 ± 0.39	0.71 ± 0.36	0.035
2nd year BCVA (logMAR)	0.99 ± 0.44	0.67 ± 0.30	0.001

Table 1. *Cont.*

Description	Group 1	Group 2	p-Value
1 year dosage	4.40 ± 1.13	4.30 ± 1.31	0.618
2 years dosage	5.43 ± 1.71	5.31 ± 1.95	0.709
HbA1c	7.56 ± 1.09	7.56 ± 1.19	0.993
NPDR/PDR	26/16	114/72	0.941
PRP	33	130	0.260
Pseudophakia	9	51	0.426
High Myopia	2	5	0.616
Smoking	1	6	1.000
Hypertension	23	70	0.302
Hyperlipidemia	4	23	0.320
Coronary artery disease	2	7	1.000
Stroke	4	5	0.108
Chronic kidney disease	9	21	0.237
Thyroid disease	0	1	1.000
Cancer	2	4	0.613

Group 1: patients with presence of OCT morphological changes after treatment; Group 2: patients without OCT morphological changes after treatment; BCVA: best corrected visual acuity; CMT: central macula thickness; PDR: proliferative diabetic retinopathy; NPDR: non-proliferative diabetic retinopathy; PRP: panretinal photocoagulation.

3.2. Subgroup Analysis for Morphological Changes

In the subgroup analysis, 38 eyes were classified into DRT group, 80 in CME group, 45 in SRD group, and 62 in VMIA group. All combinations of morphological subtypes of DME and number of conversions into another subtypes over time period of two years are summarized in Table 2. The percentage of OCT morphology changes among DRT, CME and SRD groups showed no significant difference (all $p > 0.05$). The distributions of morphological changes over the course of Ranibizumab treatment among different subtypes are shown in Figure 1. Moreover, VMIA was the most common type of change and developed in 14.9% of patients with DME receiving IVIR treatment during a follow-up period of 24 months. The number of clinically detectable ERM cases did not change significantly after the 2nd year with a mean period of 45.7 months follow-up. Overall, we found a mean period of 12.3 months when the first detection of changes between different DME subtypes.

Table 2. Combinations of morphological subtypes and changes over time period of 2 years.

OCT Patterns	All No. (%)	No. of Changes over Time Period of 2 Years (%)
DRT group		
DRT alone	38 (16.67%)	9 (23.7%)
Combined pattern	0 (0%)	0 (0%)
Total	38 (16.67%)	9 (23.7%)
CME group		
CME alone	20 (8.77%)	3 (15.0%)
Combined pattern	60 (26.3%)	14 (23.3%)
Total	80 (35.09%)	17 (21.3%)
SRD group		
SRD alone	7 (3.07%)	0 (0%)
Combined pattern	38 (16.67%)	11 (28.95%)
Total	45 (19.7%)	11(24.44%)
VMIA group		
ERM alone	13 (5.70%)	1 (7.7%)
Combined pattern	52 (22.80%)	3 (5.77%)
Total	65 (28.5%)	4 (6.15%)

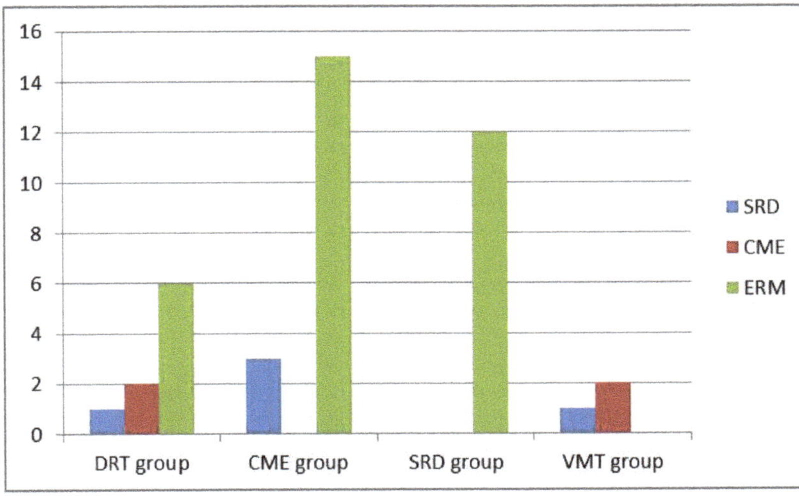

Figure 1. Distributions of morphological changes over the course of Ranibizumab treatment among different subtypes.

3.3. Correlation between Systemic Parameters and Morphology

Regarding the potential confounding factors for the morphological change in DME, stroke was significantly associated with higher risk for developing into another DME subtypes (aOR: 6.381, 95% CI: 1.112–36.623, $p = 0.038$) using multivariate logistic regression analysis. However, other demographic data and systemic parameters did not reveal significant correlation to the change in OCT morphology (all $p > 0.05$) (Table 3).

Table 3. Multivariate logistic regression analysis of morphological changes.

Variable	S.E	OR (95% CI)	*p*-Value
Intercept	1.614		0.117
Age	0.024	1.012 (0.966, 1.060)	0.615
Gender	0.440	1.425 (0.602, 3.375)	0.421
Smoking	1.242	0.487 (0.043, 5.555)	0.563
Alcohol	0.963	1.743 (0.264, 11.500)	0.564
HbA1c	0.186	1.061 (0.737, 1.526)	0.750
Hypertension	0.460	1.459 (0.593, 3.593)	0.411
CAD	0.933	0.631 (0.101, 3.925)	0.622
Stroke	0.892	6.381 (1.112, 36.623)	0.038
CKD	0.525	2.316 (0.828, 6.476)	0.109
Cancer	0.931	1.085 (0.175, 6.733)	0.930
Hyperlipidemia	0.728	0.258 (0.062, 1.075)	0.063

CAD: coronary artery disease; CKD: chronic kidney disease.

4. Discussion

In DME, liquid accumulation can occur in intracellular or extracellular spaces due to cytotoxic or vasogenic process [14]. Furthermore, pathologic fibrocellular changes at the vitreomacular interface are present in eyes with DME irrespective of the type of macular edema as classified by SD-OCT. Several patterns of macular edema depending on the location of intracellular or extracellular fluid have been first described by Otani et al. [15]. Investigators have found variable results for visual and anatomic improvement in different groups after anti-VEGF treatment [4–7,16–22]. Similarly to our current study, changes in OCT parameters over time after IVIR treatment were noted. Compared with group 2 without morphology changes, group 1 showed a significant higher CMT at the second year (p = 0.001), indicating poor response to anti-VEGF therapy and other different key mechanisms involved in the development of DME. This finding suggests that the possible strategy may shift to dexamethasone therapy or even surgical treatment.

On the aspect of VMIA groups, any alterations of OCT morphology, such as newly developed CME, SRD, or macula contracture in the development and progression of DME could occur with previous exiting vitreomacular interface abnormality. One eye with ERM further developed macula pucker at the 12th month. The CMT outcome at the second year follow-up showed an average of 390.5 ± 75.3 μm in VMIA group patients with morphological changes. The finding had considerable discrepancy in our previous work, which had favorable improvement in reduction of CMT in VMIA group with consistent IVIR injections [13]. Patients with poor response, underlying physical mechanisms of pathogenesis other than VEGF, may be involved and receive benefit from the treatment of surgical relief of traction [23]. Nevertheless, among the eyes developing ERM during the anti-VEGF treatment course, the CMT outcome at the second year showed a better resolution of 19.33 μm with shorter follow-up period compared with VMIA group, though there was no statistically significant (p = 0.62). We still assume that this difference may be caused by distinct mechanisms of ERM formation. Previous studies have stated that anti-VEGF agents may be involved in the fibrotic process of diabetic eyes, induced by the intravitreal injection method or angio-fibrotic switch of VEGF and CTGF in PDR [24]. Besides, long-term intravitreal injection attributing to repetition of volume expansion may accelerate posterior vitreous detachment, which is beneficial in the natural history of DME [25]. The ERM associated with posterior vitreous detachment may instead provide a tension force toward the retina to prevent macula edema from progression in patients whose epiretinal membranes initiated secondary to the vasculopathy process at macula, and patients with repetition of intravitreal injections. Most importantly, though CMT could achieve favorable results, symptoms of retinal distortion resulting from ERMs are still considered as a surgical indication.

Among three subtypes of DME, patients with CME or DRT had higher risk of development of VMIA than those with SRD. In accordance with the immunostaining result conducted by Haguenau, F. et al., which eyes with DRT and CME were most intensely positive for active myofibroblasts and the enzyme matrix metalloproteinase-9 (MMP-9) [26], our findings address the tendency for progression to proliferative disease in DRT and CME groups. In the present study, a succession of change was noticed from DRT to CME/SRD and from CME to SRD, although vasogenic or inflammatory etiology of DME is very difficult to distinguish from OCT. In contrary, none of eyes with pure SRD transformed into CME type after IVIR treatment, except the coexistence of initial VMIA. As vasogenic changes secondary to hyperglycemia begins the cascade of macular edema formation causing a failure in the RPE pump function [14,27,28], we suppose SRD may be an indicator of earlier stage DME, which responds well to IVIR and seldom transforms into other subtypes under intravitreal anti-VEGF therapy. Apparently, this finding correlates well with previous studies, which the presence of subretinal fluid has better response to treatment [11,29]. The individual differences of RPE function contributing to the presence of this early sign, provide in-depth explanation why part of patients developed SRD with longer DME duration and disease progression, as the schematic diagram illustrated in Figure 2.

As DM progressed, the hyperglycemia status initiates breakdown of RPE barrier. RPE cells could decompensate at any stages of DME on interindividual variation and then patients develop SRD. Consequently, accumulation of the fluid causes edema (CME/DRT). Fibrocellular proliferation on the internal limiting membrane further results in ERM secondary to DM retinopathy or intravitreal injections.

On the aspect of the association between DM retinopathy and cardiovascular diseases such as stroke or CAD, prior studies have addressed that patients with type 2 diabetes and DME or PDR have an increased risk of cardiovascular events [30]. In our present study, we found DME patient with stroke had a 6.4-fold increase in the relative risk of developing into

another DME subtypes. This finding may account for the relationship between stroke and severe diabetes, or the relevant confounding factors related to stroke. Moreover, for patients taking aspirin regularly, the anti-inflammatory agent as a confounding factor may block the gap junction communication between the RPE cells, damage retinal microenvironment [31], and possibly worsen the progression of DME.

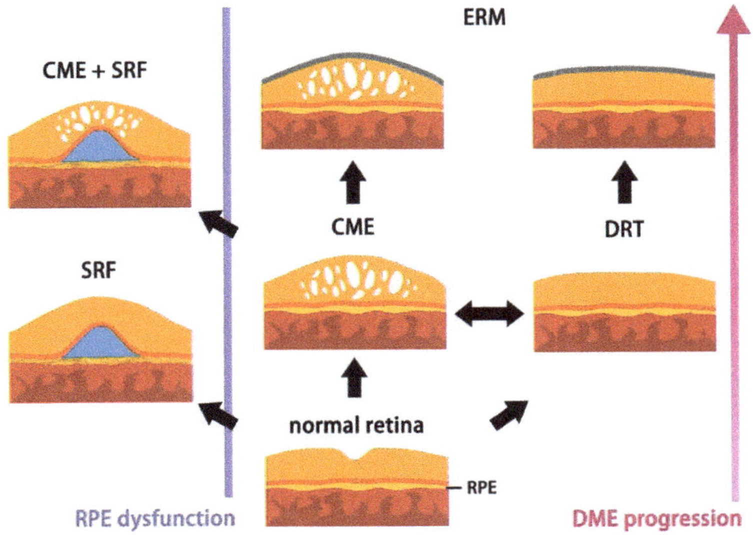

Figure 2. Schematic diagram of hypothesis of DME morphological procession changes on OCT.

Our study has several limitations to be noted. First, this nonrandomized and uncontrolled study retrospectively collected data from a relatively small number of patients without standardized treatment regimen. Second, due to the reimbursement limitations, most patients were undertreated with a mean number of 5.33 injections over 24 months. Data were derived from poor-quality images, previous treatments, or short follow-up period resulted in relative large number of excluded cases. Yet, eyes with different duration of DME, control of DM, and timing for injections were not evaluated, reflecting real-life conditions in clinical practice.

5. Conclusions

In summary, macula edema disappeared rapidly after IVIR treatment and might recur or even evolve to other subtypes over time, indicating a disease progression with poor structure and visual outcome. Pathologic fibrocellular changes at the vitreomacular interface were the most common type of changes and presented in all eyes with DME regardless of the type of macular edema as classified by SD-OCT. We suppose SRD may be an indicator of earlier stage DME, which responds well to IVIR and seldom transforms into other subtypes under intravitreal anti-VEGF therapy. Future research should focus on exploring different etiologies in DME, which would allow clinicians to offer more precise and individualized treatments.

Author Contributions: Conceptualization, C.-H.L. and N.-N.C.; methodology, C.-Y.L.; software, C.-Y.L.; validation, C.-N.K., C.-L.C., J.-C.H., P.-C.W. and C.-Y.C.; formal analysis, N.-N.C.; investigation, N.-N.C.; resources, P.-L.W.; data curation, C.-Y.C.; writing—original draft preparation, N.-N.C.; writing—review and editing, C.-Y.C.; supervision, C.-H.L. All authors have read and agreed to the published version of the manuscript.

Funding: This research received no external funding.

Institutional Review Board Statement: The current was in compliance with recognized international standards and the principles of the Declaration of Helsinki in 1964 and its late amendments. Moreover, our study was approved by the Institutional Review Board of Chang Gung Memorial Hospital (protocol code: 201901410B0, date of approval: 24 September 2019).

Informed Consent Statement: Patient consent was waived by the Institutional Review Board of Chang Gung Memorial Hospital due to the retrospective nature.

Data Availability Statement: The data related to the current study is available upon reasonable request from the editorial board.

Conflicts of Interest: The authors declare no conflict of interest.

References

1. Baskin, D.E. Optical coherence tomography in diabetic macular edema. *Curr. Opin. Ophthalmol.* **2010**, *21*, 172–177. [CrossRef] [PubMed]
2. Kim, B.Y.; Smith, S.D.; Kaiser, P.K. Optical coherence tomographic patterns of diabetic macular edema. *Am. J. Ophthalmol.* **2006**, *142*, 405–412. [CrossRef] [PubMed]
3. Cho, Y.J.; Lee, D.H.; Kim, M. Optical coherence tomography findings predictive of response to treatment in diabetic macular edema. *J. Int. Med. Res.* **2018**, *46*, 4455–4464. [CrossRef] [PubMed]
4. Buabbud, J.C.; Al-latayfeh, M.M.; Sun, J.K. Optical coherence tomography imaging for diabetic retinopathy and macular edema. *Curr. Diab. Rep.* **2010**, *10*, 264–269. [CrossRef] [PubMed]
5. Itoh, Y.; Petkovsek, D.; Kaiser, P.K.; Singh, R.P.; Ehlers, J.P. Optical coherence tomography features in diabetic macular edema and the impact on anti-vegf response. *Ophthalmic. Surg. Lasers Imaging Retin.* **2016**, *47*, 908–913. [CrossRef]
6. Cheema, H.R.; Al Habash, A.; Al-Askar, E. Improvement of visual acuity based on optical coherence tomography patterns following intravitreal bevacizumab treatment in patients with diabetic macular edema. *Int. J. Ophthalmol.* **2014**, *7*, 251–255.
7. Wu, P.C.; Lai, C.H.; Chen, C.L.; Kuo, C.N. Optical coherence tomographic patterns in diabetic macula edema can predict the effects of intravitreal bevacizumab injection as primary treatment. *J. Ocul. Pharmacol. Ther.* **2012**, *28*, 59–64. [CrossRef]
8. Alkuraya, H.; Kangave, D.; Abu El-Asrar, A.M. The correlation between optical coherence tomographic features and severity of retinopathy, macular thickness and visual acuity in diabetic macular edema. *Int. Ophthalmol.* **2005**, *26*, 93–99. [CrossRef]
9. Bressler, S.B.; Glassman, A.R.; Almukhtar, T.; Bressler, N.M.; Ferris, F.L.; Googe, J.M., Jr.; Gupta, S.K.; Jampol, L.M.; Melia, M.; Wells, J.A., 3rd; et al. Five-year outcomes of ranibizumab with prompt or deferred laser versus laser or triamcinolone plus deferred ranibizumab for diabetic macular edema. *Am. J. Ophthalmol.* **2016**, *164*, 57–68. [CrossRef]
10. Heier, J.S.; Korobelnik, J.F.; Brown, D.M.; Schmidt-Erfurth, U.; Do, D.V.; Midena, E.; Boyer, D.S.; Terasaki, H.; Kaiser, P.K.; Marcus, D.M.; et al. Intravitreal aflibercept for diabetic macular edema: 148-week results from the vista and vivid studies. *Ophthalmology* **2016**, *123*, 2376–2385. [CrossRef]
11. Sheu, S.J.; Lee, Y.Y.; Horng, Y.H.; Lin, H.S.; Lai, W.Y.; Tsen, C.L. Characteristics of diabetic macular edema on optical coherence tomography may change over time or after treatment. *Clin. Ophthalmol.* **2018**, *12*, 1887–1893. [CrossRef] [PubMed]
12. Chang, C.K.; Cheng, C.K.; Peng, C.H. The incidence and risk factors for the development of vitreomacular interface abnormality in diabetic macular edema treated with intravitreal injection of anti-vegf. *Eye* **2017**, *31*, 762–770. [CrossRef] [PubMed]
13. Chen, N.-N.; Chen, W.-D.; Lai, C.-H.; Kuo, C.-N.; Chen, C.-L.; Huang, J.-C.; Wu, P.-C.; Wu, P.-L.; Chen, C.-Y. Optical coherence tomographic patterns as predictors of structural outcome after intravitreal ranibizumab in diabetic macula edema. *Clin. Ophthalmol.* **2020**, *14*, 4023–4030. [CrossRef]
14. Romero-Aroca, P.; Baget-Bernaldiz, M.; Pareja-Rios, A.; Lopez-Galvez, M.; Navarro-Gil, R.; Verges, R. Diabetic macular edema pathophysiology: Vasogenic versus inflammatory. *J. Diabetes Res.* **2016**, *2016*, 2156273. [CrossRef] [PubMed]
15. Otani, T.; Kishi, S.; Maruyama, Y. Patterns of diabetic macular edema with optical coherence tomography. *Am. J. Ophthalmol.* **1999**, *127*, 688–693. [CrossRef]
16. Radovanova, K.T. Optical coherence tomography patterns in diabetic macular edema can predict the effectiveness of intravitreal bevacizumab combined with macular photocoagulation. *J. Clin. Exp. Ophthalmol.* **2014**, *5*. [CrossRef]
17. Hoon Seo, K.; Yu, S.-Y.; Kim, M.; Woo Kwak, H. Visual and morphologic outcomes of intravitreal ranibizumab for diabetic macular edema based on optical coherence tomography patterns. *Retina* **2016**, *36*, 588–595.
18. Kim, M.; Lee, P.; Kim, Y.; Yu, S.Y.; Kwak, H.W. Effect of intravitreal bevacizumab based on optical coherence tomography patterns of diabetic macular edema. *Ophthalmologica* **2011**, *226*, 138–144. [CrossRef]
19. Koytak, A.; Altinisik, M.; Sogutlu Sari, E.; Artunay, O.; Umurhan Akkan, J.C.; Tuncer, K. Effect of a single intravitreal bevacizumab injection on different optical coherence tomographic patterns of diabetic macular oedema. *Eye* **2013**, *27*, 716–721. [CrossRef]
20. Roh, M.I.; Kim, J.H.; Kwon, O.W. Features of optical coherence tomography are predictive of visual outcomes after intravitreal bevacizumab injection for diabetic macular edema. *Ophthalmologica* **2010**, *224*, 374–380. [CrossRef]
21. Horii, T.; Murakami, T.; Nishijima, K.; Sakamoto, A.; Ota, M.; Yoshimura, N. Optical coherence tomographic characteristics of microaneurysms in diabetic retinopathy. *Am. J. Ophthalmol.* **2010**, *150*, 840–848. [CrossRef] [PubMed]

22. Joshi, L.; Bar, A.; Tomkins-Netzer, O.; Yaganti, S.; Morarji, J.; Vouzounis, P.; Seguin-Greenstein, S.; Taylor, S.R.; Lightman, S. Intravitreal bevacizumab injections for diabetic macular edema—Predictors of response: A retrospective study. *Clin Ophthalmol.* **2016**, *10*, 2093–2098. [CrossRef] [PubMed]
23. Bonnin, S.; Sandali, O.; Bonnel, S.; Monin, C.; El Sanharawi, M. Vitrectomy with internal limiting membrane peeling for tractional and nontractional diabetic macular edema: Long-term results of a comparative study. *Retina* **2015**, *35*, 921–928. [CrossRef] [PubMed]
24. Cetin, E.N.; Demirtas, O.; Ozbakis, N.C.; Pekel, G. Quantitative assessment of macular contraction and vitreoretinal interface alterations in diabetic macular edema treated with intravitreal anti-vegf injections. *Graefes Arch. Clin. Exp. Ophthalmol.* **2018**, *256*, 1801–1806. [CrossRef]
25. Zhang, X.; Zeng, H.; Bao, S.; Wang, N.; Gillies, M.C. Diabetic macular edema: New concepts in patho-physiology and treatment. *Cell Biosci.* **2014**, *4*, 27. [CrossRef]
26. Hagenau, F.; Vogt, D.; Ziada, J.; Guenther, S.R.; Haritoglou, C.; Wolf, A.; Priglinger, S.G.; Schumann, R.G. Vitrectomy for diabetic macular edema: Optical coherence tomography criteria and pathology of the vitreomacular interface. *Am. J. Ophthalmol.* **2019**, *200*, 34–46. [CrossRef]
27. Desjardins, D.M.; Yates, P.W.; Dahrouj, M.; Liu, Y.; Crosson, C.E.; Ablonczy, Z. Progressive early breakdown of retinal pigment epithelium function in hyperglycemic rats. *Investig. Ophthalmol. Vis. Sci.* **2016**, *57*, 2706–2713. [CrossRef]
28. Daruich, A.; Matet, A.; Moulin, A.; Kowalczuk, L.; Nicolas, M.; Sellam, A.; Rothschild, P.R.; Omri, S.; Gelize, E.; Jonet, L.; et al. Mechanisms of macular edema: Beyond the surface. *Prog. Retin. Eye Res.* **2018**, *63*, 20–68. [CrossRef]
29. Sophie, R.; Lu, N.; Campochiaro, P.A. Predictors of functional and anatomic outcomes in patients with diabetic macular edema treated with ranibizumab. *Ophthalmology* **2015**, *122*, 1395–1401. [CrossRef]
30. Xie, J.; Ikram, M.K.; Cotch, M.F.; Klein, B.; Varma, R.; Shaw, J.E.; Klein, R.; Mitchell, P.; Lamoureux, E.L.; Wong, T.Y. Association of diabetic macular edema and proliferative diabetic retinopathy with cardiovascular disease: A systematic review and meta-analysis. *JAMA Ophthalmol.* **2017**, *135*, 586–593. [CrossRef]
31. Wu, Y.; Zhu, W.; Li, Y.H.; Yu, J. Aspirin and age related macular degeneration; the possible relationship. *Med. Hypothesis Discov. Innov. Ophthalmol. J.* **2013**, *2*, 59.

Journal of
*Personalized
Medicine*

MDPI

Article

Factors Predicting the Success of Combined Orbital Decompression and Strabismus Surgery in Thyroid-Associated Orbitopathy

Meng-Wei Hsieh [1,†], Chih-Kang Hsu [2,3,†], Pao-Cheng Kuo [4], Hsu-Chieh Chang [5,6], Yi-Hao Chen [3] and Ke-Hung Chien [3,*]

1 Department of Ophthalmology, Taoyuan Armed Forces General Hospital, Taoyuan 325, Taiwan; nsinkoph0311@gmail.com
2 Department of Ophthalmology, Tri-Service General Hospital Songshan Branch, Songshan, Taipei 105, Taiwan; chikanghsu@gmail.com
3 Department of Ophthalmology, Tri-Service General Hospital, National Defense Medical Center, Taipei 114, Taiwan; doc30879@mail.ndmctsgh.edu.tw
4 Department of Orthopedics, Taipei Veterans General Hospital, Taipei 112, Taiwan; king820414@gmail.com
5 Department of Nursing, Tri-Service General Hospital, National Defense Medical Center, Taipei 114, Taiwan; n3197001@gmail.com
6 Graduate Institute of Nursing, College of Nursing, Taipei Medical University, Taipei 110, Taiwan
* Correspondence: yred8530@gmail.com; Tel.: +886-2-8792-3311
† These authors contributed equally to this work.

Abstract: To evaluate the safety and efficacy of orbital decompression combined with strabismus surgery in thyroid-associated orbitopathy (TAO) and identify factors leading to surgical success. A retrospective comparative case series was conducted on 52 patients who were treated with combined orbital decompression and strabismus surgery. Outcome measurements included perioperative Hertel exophthalmometry and strabismus measurements. Surgical success was defined as binocular single vision (BSV) in the primary and reading positions within 5 prism diopters (PDs). As a result, the average reduction in proptosis was 3.23 mm, with a mean preoperative Hertel measurement of 22.64 mm. Forty-four patients (84.6%) achieved the success criterion and composed the success group. In addition to sex and underlying hyperthyroidism, symmetry of orbitopathy, interocular exophthalmos difference of more than 2 mm, predominant esotropia type, mixed type strabismus, baseline horizontal deviations, baseline vertical deviations, and combination with one-wall decompression surgery were significantly different between the success and failure groups. All complications were mild and temporary. Orbital decompression combined with strabismus surgery produced satisfactory outcomes in selected patients with efficacy and safety. Symmetry between the two eyes with relatively simple strabismus and proptosis ensured surgical success. With experienced surgeons, advanced techniques, and selected patients, this method can serve as an alternative treatment option to minimize the number of surgeries, medical costs and recovery period.

Keywords: hyperthyroidism; strabismus; orbital decompression surgery

Citation: Hsieh, M.-W.; Hsu, C.-K.; Kuo, P.-C.; Chang, H.-C.; Chen, Y.-H.; Chien, K.-H. Factors Predicting the Success of Combined Orbital Decompression and Strabismus Surgery in Thyroid-Associated Orbitopathy. *J. Pers. Med.* **2022**, *12*, 186. https://doi.org/10.3390/jpm12020186

Academic Editor: Chieh-Chih Tsai

Received: 4 December 2021
Accepted: 27 January 2022
Published: 31 January 2022

1. Introduction

Thyroid-associated orbitopathy (TAO) is a complex disorder associated with autoimmunity against orbital tissues and is characterized by inflammation of retrobulbar tissues, adipogenesis, and the accumulation of glycosaminoglycans within the extraocular muscles [1]. Clinical presentations include proptosis, diplopia, lid retraction, corneal erosion, and visual impairment [2]. The symptoms could be highly variable even between the two eyes of the same patient [3]. Medication is the first line of treatment [4]. However, up to 20% of TAO patients undergo surgical intervention [5].

In 1986, Shorr and Seiff [6] proposed the staged approach in the surgical treatment of TAO patients: (1) orbital decompression, (2) extraocular muscle surgery, (3) eyelid margin repositioning, and (4) blepharoplasty. This approach has become a consensus and has been carried out for most TAO patients requiring surgical intervention. Orbital decompression is often suggested first, as it can cause ocular alignment shifting, affecting strabismic measurements and eyelid positioning [7,8]. Decompression-related disease reactivation is rare but also a concern [9]. Nonetheless, a wide variety of surgical techniques have been introduced and improved over the years, and the rate of patients undergoing orbital decompression for TAO has increased substantially [10]. However, it has been changed since the introduction of Tepezza [11]. Current indications for surgery have also become broad, from cases of optic neuropathy to extreme corneal exposure to cosmetic rejuvenation, and surgical techniques are often individualized for different cases [12–14].

The first attempt at simultaneous orbital decompression and eye muscle surgery was described by Michel et al. in 2001 to correct the alignment shifting caused by medial wall decompression. Ocular motility was corrected by medial rectus muscle recession in 58 patients. They achieved free of double vision within margins of 20 degrees in each direction in 31 out of 58 patients after the initial operation [15]. Moreover, studies combining procedures in different stages have emerged in recent years, including orbital decompression surgery combined with eyelid or strabismus surgery. Most studies have shown a reduced number of surgeries without compromising results compared with the staged process [16–20].

In our study, the medical records of 52 TAO patients were reviewed, and all patients underwent combined orbital decompression and strabismus surgery. They were then divided into a success group (44 patients) and a failure group (8 patients). We reviewed the surgical outcomes, the preoperative and postoperative data, and the events of complications and compared the results with those of previous studies. The goal of this study was to evaluate and provide another perspective of the surgical approach to TAO using a combined surgical method for orbital decompression and strabismus correction.

2. Materials and Methods

The medical records of patients undergoing combined orbital decompression and strabismus surgery at Tri-Service General Hospital for mild to moderate TAO from January 2015 to December 2020 were retrospectively reviewed. The study was approved by the Institutional Review Board of Tri-Service General Hospital (1-107-05-119).

All patients with hyperthyroidism had received medication treatment at the Endocrinology Department first, and stable disease with euthyroid status was achieved. Data was collected included age, sex, clinical manifestations of TAO, past medical history, and smoking status. Preoperative and postoperative evaluations included visual acuity (VA), intraocular pressure (IOP), a slit lamp exam, Hertel exophthalmometry, extraocular movements (EOMs), and prism tests. A preoperative imaging study was performed for every patient with computed tomography scans (CT) to assess orbital crowding and muscle hypertrophy and aid in surgical planning.

The indications of the combined surgery were as follows: (1) diagnosis of thyroid-associated orbitopathy; (2) disfiguring exophthalmos requiring cosmetic rehabilitation; (3) restrictive strabismus associated with extraocular muscle hypertrophy and symptoms of diplopia; and (4) inactive disease (clinical activity score less than 3) with euthyroid state. Surgery was performed after both euthyroid state after medication control and stable ophthalmic signs with regular follow-ups for more than 6 months were achieved.

All surgeries were performed under general anesthesia by one surgeon (Ke-Hung Chien, MD). A transconjunctival approach was employed for orbital decompression, and adjunctive fat decompression was performed simultaneously if needed. The extent of bone and fat removal was individualized according to the proptosis severity and followed the TAO management protocol [21]. The protocol was mainly determined by Hertel exophthalmeter measurements and the presence of diplopia. Patients were selected for combined surgeries in this study if they had ocular deviation of more than 10 PD in either horizontal or vertical dimension. Strabismus surgery was performed in the same session after orbital decompression, and extraocular muscle recession and/or resection was performed according to preoperative measurements of restrictive strabismus.

Some modifications were made to prevent complications:

(1) We created two separate incisions for different procedures. For example, one inferomedial conjunctival forniceal incision was used for inferomedial orbital decompression, and one inferolateral conjunctival forniceal incision was used for inferior rectus muscle surgery, and inferomedial conjunctival limbal incision was used for medial rectus muscle surgery. Usually, recession of extraocular muscles was done, and resection was performed occasionally to further correct strabismus.
(2) During inferomedial orbital decompression, we performed strut preservation to minimize surgery-induced diplopia and consecutive hypoglobus [22].
(3) During strabismus surgery, we followed the surgical dose response modification with the "intraoperative relaxed muscle positioning technique" [23–26].

Follow-up was arranged at one week, one month, and three months postoperatively. The clinical outcome measures included Hertel exophthalmometry and strabismus measurements.

The success criteria were set from the literature and modified according to clinical practice [27–29]. We defined success in our study as binocular single vision (BSV) in the primary and reading positions within 5 prism diopters (PDs) at least 3 months after the first operation. The criteria of failure were met if the patient had persistent diplopia, had correctable diplopia with more than 5 PDs in the primary and reading positions, or underwent a second operation within 3 months after the first operation.

We used SPSS statistical software version 18 (SPSS Inc, IBM Company, Chicago, IL, USA) to perform statistical analyses. Continuous variables between group 1 and group 2 were analyzed by *t*-test. The variables for the success factors analysis were identified using univariate and multivariate linear regression analyses. A p value < 0.05 was considered statistically significant.

3. Results

A total of 52 subjects underwent combined orbital decompression and strabismus surgery at Tri-Service General Hospital and were included in the study.

The whole group comprised 35 females and 17 males, with a mean age of 56.07 years ranging from 29.2 to 91.6 years. There were 45 patients who developed TAO with underlying hyperthyroidism, with 3 cases of Hashimoto's thyroiditis, 2 cases of hypothyroidism, and 2 cases of consecutive hypothyroidism (from thyroidectomy). Mean follow-up time of the patients was 24.7 months, ranging from 3.6 months to 132.7 months (Table 1).

Table 1. Demographic characteristics of the patients in the study.

	Whole Group	Group 1 (Success)	Group 2 (Failure)	p Value
No. of subjects (N) (%)	52 (100%)	44 (84.6%)	8 (15.4%)	
Male (N) (%)	17 (32.7%)	15 (34.1%)	2 (25.0%)	0.04 *
Female (N) (%)	35 (67.3%)	29 (65.9%)	6 (75.0%)	
Age at operation (years) (mean) (SD)	56.07 (14.18)	54.95 (13.15)	57.29 (16.38)	0.66
Age at operation (years) (min) (max)	(29.2) (91.6)	(29.2) (77.8)	(36.1) (91.6)	
Follow-up period (month) (mean) (SD)	24.7 (27.9)	28.2 (23.9)	16.44 (8.23)	0.17
Follow-up period (min) (max)	(3.6) (132.7)	(3.6) (132.7)	(4.2) (70.5)	
Underlying hyperthyroidism (N) (%)	45 (86.5%)	39 (88.6%)	6 (75%)	0.03 *

The p value in the table is from values compared between Group 1 and Group 2. N = case number. SD = standard deviation. $p < 0.05$ = significant (*).

In reviewing the patients' background prior to the surgery, 24 patients (46.2%) had a history of smoking, including active status and quitting status after TAO diagnosis. All patients had been receiving medical treatment and attained the euthyroid state at least 6 months before surgery, 33 patients (63.5%) had a history of prior usage of corticosteroids, 7 patients (13.5%) underwent radioactive iodine treatment, and 1 patient (1.9%) underwent radiotherapy. The mean duration of TAO was 29.2 months (Table 2).

Table 2. Background of patients prior to the surgery.

	Whole Group (N = 52)	Group 1 (Success) (N = 44)	Group 2 (Failure) (N = 8)	*p* Value
Relative symmetry of orbitopathy (Less than 2 mm difference) (N) (%)	47 (90.4%)	42 (93.2%)	5 (62.5%)	0.02 *
Preoperative exophthalmos difference (mm) (mean) (SD)	1.23 (0.86)	1.08 (0.81)	2.05 (0.94)	0.03 *
Preoperative exophthalmos (mm) (mean) (SD)	22.64 (2.56)	22.36 (2.41)	24.18 (4.12)	0.63
Postoperative exophthalmos (mm) (mean) (SD)	19.41 (1.73)	19.24 (1.65)	20.35 (2.13)	0.39
Smoking including active and quit (N) (%)	24 (46.2%)	21 (47.8%)	3 (37.5%)	0.05
Prior corticosteroid usage (N) (%)	33 (63.5%)	28 (63.6%)	5 (62.5%)	0.11
Prior radioactive iodine usage (N) (%)	7 (13.5%)	6 (13.6%)	1 (12.5%)	0.13
Prior radiotherapy (N) (%)	1 (1.9%)	1 (2.3%)	0 (0)	0.06
TAO duration (month) (mean) (SD)	29.27 (13.1)	29.1 (12.3)	30.2 (14.5)	0.15

The *p* value in the table is from values compared between Group 1 and Group 2. N = case number. SD = standard deviation. TAO = thyroid-associated ophthalmopathy. $p < 0.05$ = significant (*).

In the whole group, the patients had a preoperative exophthalmos of 22.64 mm (SD = 3.62) and a postoperative exophthalmos of 19.41 mm (SD = 1.82). Assessing the exophthalmos difference between the two eyes, 47 patients (90.4%) had a relative symmetry of their orbitopathy (interocular exophthalmos difference less than 2 mm). Strabismus exams showed 8 patients (15.4%) with predominant esotropia (with a vertical component less than 5 PDs), 6 patients (11.5%) with predominant hypotropia (with a horizontal component less than 5 PDs), and 38 patients (73.1%) with mixed strabismus. Their mean baseline horizontal deviation was 27.6 PDs (SD = 8.6), and the mean baseline vertical deviation was 9.1 PDs (SD = 6.2). Additionally, 20 patients (38.5%) were found to have cyclotorsion before surgery (Table 3).

Table 3. Perioperative examinations of the patients in the study.

	Whole Group (N = 52)	Group 1 (Success) (N = 44)	Group 2 (Failure) (N = 8)	*p* Value
Type of strabismus				
Predominantly esotropia (N) (%) (Without other component > 5 PDs)	8 (15.4%)	8 (18.2%)	0 (0)	0.03 *
Predominantly hypotropia (N) (%) (Without other component > 5 PDs)	6 (11.5%)	5 (11.4%)	1 (12.5%)	0.17
Mixed type strabismus (N) (%) (More than one component > 5 PDs)	38 (73.1%)	31 (70.5%)	7 (87.5%)	0.02 *
Baseline horizontal deviations [1] (PD) (mean) (SD)	27.6 (8.8)	26.4 (8.5)	34.1 (9.7)	0.03 *
Baseline vertical deviations [1] (PD) (mean) (SD)	9.1 (7.0)	8.3 (6.7)	13.6 (7.6)	0.02 *
Cyclotorsion (N) (%)	20 (38.5%)	16 (36.4%)	4 (50%)	0.66
Combination surgery with one muscle (N) (%)	16 (30.8%)	14 (31.8%)	2 (25%)	0.33
Combination surgery with two or more muscles (N) (%)	36 (69.2%)	30 (68.2%)	6 (75%)	0.41
Combination with one-wall decompression (N) (%)	27 (51.9%)	26 (59.1%)	1 (12.5%)	0.02 *
Combination with two or more walls decompression (N) (%)	25 (48.1%)	18 (40.9%)	7 (87.5%)	0.03 *

The *p* value in the table is from values compared between Group 1 and Group 2. N = case number. SD = standard deviation. PD = prism diopter. $p < 0.05$ = significant (*). 1 Deviation of esotropia and hypotropia were in plus numbers while deviation in exotropia and hypertropia were in minus numbers.

Generally, we followed the TAO management protocol [21] with some modifications (detailed in the Methods section). In total, 16 patients (30.8%) underwent single muscle surgery, and 27 patients (51.9%) underwent one-wall decompression in their combination surgery.

To further evaluate the factors in combination surgery, the whole group was then divided into two groups (the success and failure groups) according to their postoperative strabismus status, with a criterion of BSV in the primary and reading positions within 5 PDs at the 3-month follow-up visit. Following the criteria, 44 patients (84.62%) were included in the success group, while 8 (15.38%) were included in the failure group.

Regarding the characteristics of the subgroups, there were significant differences in sex ($p = 0.04$) and underlying hyperthyroidism ($p = 0.03$) (Table 1). In the baseline TAO examinations, symmetry of orbitopathy ($p = 0.02$) and interocular exophthalmos difference more than 2 mm ($p = 0.03$) were significantly different between the groups, while there was no significant difference in other parameters, such as preoperative exophthalmos measurement, smoking status, history of corticosteroid usage, radioactive iodine, radiotherapy, and duration of TAO (Table 2). In the evaluation of operative parameters, predominant esotropia type, mixed-type strabismus, baseline horizontal deviations, baseline vertical deviations, and combination with one-wall decompression surgery were significantly different between the two groups (Table 3).

To determine which parameter most commonly leads to the success of combination surgery, we performed statistical analysis on the relationships among all baseline characteristics and preoperative ophthalmic exams with surgical success. According to the results, symmetry of orbitopathy ($p < 0.001$), interocular exophthalmos difference of more than 2 mm ($p = 0.012$), mixed-type strabismus ($p = 0.041$), baseline vertical deviation ($p = 0.011$) and combination with one-wall decompression surgery ($p = 0.036$) were significantly related to surgical success (all with negative correlations). According to the correlation statistics, interocular exophthalmos difference of greater than 2 mm ($r = -0.412$), mixed-type strabismus ($r = -0.127$), and baseline vertical deviation ($r = -0.438$) had negative correlations, while symmetry of orbitopathy ($r = 0.541$) and combination with one-wall decompression surgery ($r = 0.273$) had positive correlations. This result implied that patients with more symmetry between the two orbits and less proptosis more easily tended to experience surgical success in the combination surgery.

There were no major complications noted in our cohort, such as cerebrospinal fluid (CSF) leakage, global injury, or vision loss. However, there were five patients (9.6%) (three patients (6.8%) in the success group and two patients (25%) in the failure group) with temporary infraorbital nerve compromise. One patient in the success group experienced postoperative soft tissue adhesion (symblepharon).

4. Discussion

Current minimally invasive surgical decompression techniques have improved surgical outcomes and demonstrated potential in combination with additional procedures [30]. Combination surgeries for TAO management, including two or three of the staged surgeries, are now gaining popularity. In our study, strabismus correction was performed by recessing the hypertrophied extraocular muscles after orbital decompression in the same session. The results showed an 84.6% success rate with the criterion of any postoperative deviations less than 5 PDs. To the best of our knowledge, in addition to our study, there was a study published in 2016 by Seung Woo Choi [18] that reported five patients who underwent combined orbital decompression and strabismus surgery. The study included three cases of lateral wall decompression, one case of balanced decompression, and one case of fat decompression. The mean postoperative Hertel value reduction was 2.1 ± 2.0 mm. The sum of the angles in BSV for the five patients who underwent combined surgery improved from 101.3° to 318.8° ($p = 0.068$). Our study included 52 patients and showed a comparable reduction in exophthalmos measurement (3.23 ± 2.21 mm in the whole group, 3.12 ± 1.73 mm in the success group, and 3.83 ± 3.12 mm in the failure group), even when combined with strabismus surgery. This finding demonstrated that combination surgery could be an effective treatment choice in these patients.

Inferomedial decompression by itself can have widely variable effects, depending on the surgeon technique and extent of bony decompression. For the transnasal and transorbital approaches, studies have shown postoperative diplopia rates ranging from 10% to 35% [31–34]. The technique was adopted in our study mainly due to surgeon preference and followed the modification to minimize surgically induced diplopia. Modifications of decompression techniques, including inferomedial strut preservation and selective periosteal dissection, have been reported to reduce the incidence of postoperative diplopia [22,35–37]. According to our experience, a limited range of decompression with preservation of vital structures can produce a more stable and predictable surgical outcome.

Some studies have reported that lateral wall decompression shows preferable results, with new-onset diplopia rates ranging from 0% to 6% [38–43]. This method was adopted in the combined surgery performed in the study of Seung Woo Choi [18], which also produced results comparable to those of staged surgery. However, orbital decompression techniques still lack consistency in outcome, and nonuniform strabismus measurements also limit the ability to make evidence-based decisions [44]. Thus, in regard to practice, surgeon experience and preference still play a major role in adopting decompression techniques. In our study, lateral wall decompression was only considered in patients with a preoperative exophthalmos greater than 24 mm in either balanced decompression or three-wall decompression [21]. The protocol showed reliable results even when combined with strabismus correction in our study.

Studies have reported that 17–45% of TAO patients require more than one procedure to correct TAO-related strabismus [45]. It often causes significant difficulty in daily activities, and early correction may greatly improve quality of life [25,46]. In addition, strabismus surgery is difficult to concur due to significant postoperative drift and different surgical dose responses with common strabismus surgeries [23,24,27,47]. To more precisely evaluate the strabismus result, we modified the intraoperative relaxed muscle positioning technique in which the extent of recession was mainly determined by the position in which the muscle tendon rests freely on the eyeball while in the primary position [25,26]. The technique ensured our surgical expectations during combination surgery and alleviated both the surgeon's and patient's burden.

The most alarming complication in the combination surgery is symblepharon from soft tissue adhesion between different surgical planes, which may result in orbital adhesion syndrome. We modified the surgical planes in one session by creating two surgical incisions (one for strabismus surgery and another for orbital decompression) and performing one procedure after closing the other surgical incision. With this method, we encountered only one symblepharon case (1.9%) in the whole group. The symblepharon was relieved by a simple cutting in OPD. There was no abnormality of extraocular movement noted after the symblepharon was relieved.

There were two limitations in our study. First, the study was performed in a retrospective design and lacked a control group. Since staged TAO management is seen as a traditional treatment with proven efficacy, the purpose of our study was to emphasize the relative efficacy of a one-stage operation. We focused on comparing the characteristics between the two subgroups to clarify the factors leading to success. Based on our study results, a further prospective study should be performed to support the benefits of combination surgery. Second, there was only one surgeon involved in the study, and the procedures in the combination surgery were mainly decided by the surgeon's preference. To minimize the bias of the surgeon's preference in surgical choice, we followed the TAO management protocol with modifications in the decompression and strabismus techniques. We believe these modifications made our study more practical and reliable in real practice.

5. Conclusions

In conclusion, our study shows similar outcomes to those of a previous study performed by Seung Woo Choi [18], which yielded comparable outcomes with the staged process. While there was still a limited number of patients included in our study and more well-designed studies are needed to analyze the safety and effectiveness of the procedure, we believe that with experienced surgeons and advanced techniques, combination surgery can serve as an alternative treatment option to minimize the number of surgeries, medical costs, and recovery period in selected patients.

Author Contributions: Design of the study (K.-H.C. and C.-K.H.) and conduct of the study (K.-H.C. and P.-C.K.); data collection (H.-C.C.), data analysis (M.-W.H.), and interpretation of the data (K.-H.C. and H.-C.C.); and preparation (K.-H.C. and M.-W.H.), review (K.-H.C.), and approval of the manuscript (Y.-H.C.). All authors have read and agreed to the published version of the manuscript.

Funding: This research was funded by the TSGH (Tri-Service General Hospital) (TSGH-E-109231, and TSGH-D-110115), Tri-Service General Hospital Songshan branch (TSGH-SS-D-109020, and TSGH-SS-A-110001) and TSGH-TAFGH (Taoyuan Armed Forces General Hospital) Joint Research Program (TYAFGH-A-110019, TYAFGH-E-111044), the Ministry of National Defense Medical Affairs Bureau (MND-MAB-C05-111019), and the Taiwan Ministry of Science and Technology (MOST 110-2314-B-016-051).

Institutional Review Board Statement: The study was conducted according to the guidelines of the Declaration of Helsinki and approved by the Ethics Committee of Tri-Service General Hospital (1-107-05-119; Date: 25 August 2018).

Data Availability Statement: The datasets used and analyzed during the current study are available from the corresponding author on reasonable request.

Conflicts of Interest: The authors declare no conflict of interest.

References

1. Salvi, M.; Campi, I. Medical Treatment of Graves' Orbitopathy. *Horm. Metab. Res.* **2015**, *47*, 779–788. [CrossRef] [PubMed]
2. Weetman, A.P. Thyroid-associated eye disease: Pathophysiology. *Lancet* **1991**, *338*, 25–28. [CrossRef]
3. Chiarelli, A.G.M.; De Min, V.; Saetti, R.; Fusetti, S.; Al Barbir, H. Surgical management of thyroid orbitopathy. *J. Plast. Reconstr. Aesthet. Surg.* **2010**, *63*, 240–246. [CrossRef] [PubMed]
4. Baldeschi, L.; Wakelkamp, I.M.M.J.; Lindeboom, R.; Prummel, M.F.; Wiersinga, W.M. Early versus Late Orbital Decompression in Graves' Orbitopathy. A Retrospective Study in 125 Patients. *Ophthalmology* **2006**, *113*, 874–878. [CrossRef] [PubMed]
5. Bartley, G.B.; Fatourechi, V.; Kadrmas, E.F.; Jacobsen, S.J.; Ilstrup, D.M.; Garrity, J.A.; Gorman, C.A. The treatment of Graves' ophthalmopathy in an incidence cohort. *Am. J. Ophthalmol.* **1996**, *121*, 200–206. [CrossRef]
6. Shorr, N.; Seiff, S.R. The Four Stages of Surgical Rehabilitation of the Patient with Dysthyroid Ophthalmopathy. *Ophthalmology* **1986**, *93*, 476–483. [CrossRef]
7. Rootman, D.B.; Golan, S.; Pavlovich, P.; Rootman, J. Postoperative Changes in Strabismus, Ductions, Exophthalmometry, and Eyelid Retraction after Orbital Decompression for Thyroid Orbitopathy. *Ophthalmic Plast. Reconstr. Surg.* **2017**, *33*, 289–293. [CrossRef]
8. Santos De Souza Lima, L.C.; Velarde, L.G.C.; Vianna, R.N.G.; Herzog Neto, G. The effect of horizontal strabismus surgery on the vertical palpebral fissure width. *J. AAPOS* **2011**, *15*, 473–475. [CrossRef]
9. Baldeschi, L.; Lupetti, A.; Vu, P.; Wakelkamp, I.M.M.J.; Prummel, M.F.; Wiersinga, W.M. Reactivation of Graves' Orbitopathy after Rehabilitative Orbital Decompression. *Ophthalmology* **2007**, *114*, 1395–1402. [CrossRef]
10. Dharmasena, A.; Keenan, T.D.L.; Goldacre, M.J. Orbital decompression for thyroid-associated orbitopathy in England: Trends over time and geographical variation. *Orbit* **2014**, *33*, 109–114. [CrossRef]
11. Patel, A.; Yang, H.; Douglas, R.S. A New Era in the Treatment of Thyroid Eye Disease. *Am. J. Ophthalmol.* **2019**, *208*, 281–288. [CrossRef] [PubMed]
12. Lyons, C.J.; Rootman, J. Orbital Decompression for Disfiguring Exophthalmos in Thyroid Orbitopathy. *Ophthalmology* **1994**, *101*, 223–230. [CrossRef]
13. Gorman, C.A.; Desanto, L.W.; Maccarty, C.S.; Riley, F.C. Optic Neuropathy of Graves's Disease: Treatment by Transantral or Transfrontal Orbital Decompression. *N. Engl. J. Med.* **1974**, *290*, 70–75. [CrossRef] [PubMed]
14. Kikkawa, D.O.; Pornpanich, K.; Cruz, R.C.; Levi, L.; Granet, D.B. Graded orbital decompression based on severity of proptosis. *Ophthalmology* **2002**, *109*, 1219–1224. [CrossRef]
15. Michel, O.; Oberländer, N.; Neugebauer, P.; Neugebauer, A.; Rüßmann, W. Follow-up of transnasal orbital decompression in severe Graves' ophthalmopathy. *Ophthalmology* **2001**, *108*, 400–404. [CrossRef]

16. Bernardini, F.P.; Skippen, B.; Zambelli, A.; Riesco, B.; Devoto, M.H. Simultaneous Aesthetic Eyelid Surgery and Orbital Decompression for Rehabilitation of Thyroid Eye Disease: The One-Stage Approach. *Aesthet. Surg. J.* **2018**, *38*, 1052–1061. [CrossRef]
17. Ben Simon, G.J.; Mansury, A.M.; Schwarcz, R.M.; Lee, S.; McCann, J.D.; Goldberg, R.A. Simultaneous orbital decompression and correction of upper eyelid retraction versus staged procedures in thyroid-related orbitopathy. *Ophthalmology* **2005**, *112*, 923–932. [CrossRef]
18. Choi, S.W.; Lee, J.Y.; Lew, H. Customized Orbital Decompression Surgery Combined with Eyelid Surgery or Strabismus Surgery in Mild to Moderate Thyroid-associated Ophthalmopathy. *Korean J. Ophthalmol.* **2016**, *30*, 1–9. [CrossRef]
19. Taban, M.R. Combined orbital decompression and lower eyelid retraction surgery. *J. Curr. Ophthalmol.* **2018**, *30*, 169–173. [CrossRef] [PubMed]
20. Norris, J.H.; Ross, J.J.; O'Reilly, P.; Malhotra, R. A review of combined orbital decompression and lower eyelid recession surgery for lower eyelid retraction in thyroid orbitopathy. *Br. J. Ophthalmol.* **2011**, *95*, 1664–1669. [CrossRef]
21. Cheng, A.M.S.; Wei, Y.H.; Liao, S.L. Strategies in Surgical Decompression for Thyroid Eye Disease. *Oxidative Med. Cell. Longev.* **2020**, *2020*, 3537675. [CrossRef] [PubMed]
22. Endoscopic Orbital Decompression with Preservation of an Inferomedial Bony Strut: Minimization of Postoperative Diplopia—PubMed. Available online: https://pubmed.ncbi.nlm.nih.gov/10579153/ (accessed on 4 June 2021).
23. Akbari, M.; Bayat, R.; Mirmohammadsadeghi, A.; Mahmoudzadeh, R.; Eshraghi, B.; Salabati, M. Strabismus Surgery in Thyroid-Associated Ophthalmopathy; Surgical Outcomes and Surgical Dose Responses. *J. Binocul. Vis. Ocul. Motil.* **2020**, *70*, 150–156. [CrossRef] [PubMed]
24. Honglertnapakul, W.; Cavuoto, K.M.; McKeown, C.A.; Capó, H. Surgical treatment of strabismus in thyroid eye disease: Characteristics, dose–response, and outcomes. *J. AAPOS* **2020**, *24*, 72.e1–72.e7. [CrossRef] [PubMed]
25. Dal Canto, A.J.; Crowe, S.; Perry, J.D.; Traboulsi, E.I. Intraoperative Relaxed Muscle Positioning Technique for Strabismus Repair in Thyroid Eye Disease. *Ophthalmology* **2006**, *113*, 2324–2330. [CrossRef] [PubMed]
26. Sarici, A.M.; Mergen, B.; Oguz, V.; Dogan, C. Intraoperative relaxed muscle positioning technique results in a tertiary Center for Thyroid Orbitopathy Related Strabismus 11 Medical and Health Sciences 1103 Clinical Sciences. *BMC Ophthalmol.* **2018**, *18*, 305. [CrossRef]
27. Nassar, M.M.; Dickinson, A.J.; Neoh, C.; Powell, C.; Buck, D.; Galal, E.; Clarke, M.P. Parameters predicting outcomes of strabismus surgery in the management of Graves' ophthalmopathy. *J. AAPOS* **2009**, *13*, 236–240. [CrossRef] [PubMed]
28. Schittkowski, M.; Fichter, N.; Guthoff, R. Strabismus surgery in Grave's disease–dose-effect relationships and functional results. *Klinische Monatsblatter fur Augenheilkunde* **2004**, *221*, 941–947. [CrossRef]
29. Pitchon, E.M.; Klainguti, G. Surgical treatment of diplopia in Graves' orbitopathy. *Klinische Monatsblatter fur Augenheilkunde* **2007**, *224*, 331–333. [CrossRef]
30. Douglas, R.S. Commentary on: Simultaneous Aesthetic Eyelid Surgery and Orbital Decompression for Rehabilitation of Thyroid Eye Disease: The One-Stage Approach. *Aesthet. Surg. J.* **2018**, *38*, 1062–1064. [CrossRef]
31. Borumandi, F.; Hammer, B.; Kamer, L.; Von Arx, G. How predictable is exophthalmos reduction in Graves' orbitopathy? A review of the literature. *Br. J. Ophthalmol.* **2011**, *95*, 1625–1630. [CrossRef]
32. Mourits, M.P.; Bijl, H.; Altea, M.A.; Baldeschi, L.; Boboridis, K.; Currò, N.; Dickinson, A.J.; Eckstein, A.; Freidel, M. Outcome of orbital decompression for disfiguring proptosis in patients with Graves' orbitopathy using various surgical procedures. *Br. J. Ophthalmol.* **2009**, *93*, 1518–1523. [CrossRef] [PubMed]
33. Leong, S.C.; Karkos, P.D.; MacEwen, C.J. White PS. A systematic review of outcomes following surgical decompression for dysthyroid orbitopathy. *Laryngoscope* **2009**, *119*, 1106–1115. [CrossRef] [PubMed]
34. Paridaens, D.; Lie, A.; Grootendorst, R.J.; van den Bosch, W.A. Efficacy and side effects of "swinging eyelid" orbital decompression in Graves' orbitopathy: A proposal for standardized evaluation of diplopia. *Eye* **2006**, *20*, 154–162. [CrossRef] [PubMed]
35. Jordan, D.R.; Anderson, R.L. Orbital decompression. *Ophthalmic Plast. Reconstr. Surg.* **2000**, *16*, 167–168. [CrossRef]
36. Jimenez-Chobillon, M.A.; Lopez-Oliver, R.D. Transnasal endoscopic approach in the treatment of Graves ophthalmopathy: The value of a medial periorbital strip. *Eur. Ann. Otorhinolaryngol. Head Neck Dis.* **2010**, *127*, 97–103. [CrossRef]
37. Metson, R.; Samaha, M. Reduction of diplopia following endoscopic orbital decompression: The orbital sling technique. *Laryngoscope* **2002**, *112*, 1753–1757. [CrossRef]
38. Ben Simon, G.J.; Syed, A.M.; Lee, S.; Wang, D.Y.; Schwarcz, R.M.; McCann, J.D.; Goldberg, R.A. Strabismus after Deep Lateral Wall Orbital Decompression in Thyroid-Related Orbitopathy Patients Using Automated Hess Screen. *Ophthalmology* **2006**, *113*, 1050–1055. [CrossRef]
39. Ben Simon, G.J.; Wang, L.; McCann, J.D.; Goldberg, R.A. Primary-gaze diplopia in patients with thyroid-related orbitopathy undergoing deep lateral orbital decompression with intraconal fat debulking: A retrospective analysis of treatment outcome. *Thyroid* **2004**, *14*, 379–383. [CrossRef]
40. Chang, E.L.; Piva, A.P. Temporal Fossa Orbital Decompression for Treatment of Disfiguring Thyroid-Related Orbitopathy. *Ophthalmology* **2008**, *115*, 1613–1619. [CrossRef]
41. Korinth, M.C.; Ince, A.; Banghard, W.; Gilsbach, J.M. Follow-up of extended pterional orbital decompression in severe Graves' ophthalmopathy. *Acta Neurochir.* **2002**, *144*, 113–120. [CrossRef]

42. Liao, S.L.; Shih, M.J.; Chang, T.C.; Lin, L.L.K. Transforniceal lateral deep bone decompression—A modified techique to prevent postoperative diplopia in patients with disfiguring exophthalmos due to dysthyroid orbitopathy. *J. Formos. Med. Assoc.* **2006**, *105*, 611–616. [CrossRef]
43. Nguyen, J.; Fay, A.; Yadav, P.; MacIntosh, P.W.; Metson, R. Stereotactic microdebrider in deep lateral orbital decompression for patients with thyroid eye disease. *Ophthalmic Plast. Reconstr. Surg.* **2014**, *30*, 262–266. [CrossRef] [PubMed]
44. Rootman, D.B. Orbital decompression for thyroid eye disease. *Surv. Ophthalmol.* **2018**, *63*, 86–104. [CrossRef]
45. Volpe, N.J.; Mirza-George, N.; Binenbaum, G. Surgical management of vertical ocular misalignment in thyroid eye disease using an adjustable suture technique. *J. AAPOS* **2012**, *16*, 518–522. [CrossRef] [PubMed]
46. Schaut, A.; Cloché, V.; Mouna, A.; Angioi, K.; Berrod, J.P.; Conart, J.B.; Maalouf, T. Evaluation by quality of life questionnaires in patients undergoing strabismus surgery in Graves' disease. *J. Fr. D'ophtalmol.* **2018**, *41*, 814–822. [CrossRef]
47. Garrity, J.A.; Greninger, D.A.; Ekdawi, N.S.; Steele, E.A. The management of large-angle esotropia in Graves ophthalmopathy with combined medial rectus recession and lateral rectus resection. *J. AAPOS* **2019**, *23*, 15.e1–15.e5. [CrossRef]

Journal of *Personalized Medicine*

Article

Management of Delayed Complications of Hydrogel Scleral Buckles

Hsin-Yu Yang [1,2,3], Wei-Kuang Yu [2,3] and Chieh-Chih Tsai [2,3,*]

1 Taipei Veterans General Hospital Yuanshan and Suao Branch, Yilan 264018, Taiwan; lisa771102@yahoo.com.tw
2 Department of Ophthalmology, Taipei Veterans General Hospital, Taipei 11217, Taiwan; hikaruyu@gmail.com
3 School of Medicine, National Yang Ming Chiao Tung University, Taipei 11221, Taiwan
* Correspondence: cctsai@vghtpe.gov.tw; Tel.: +886-2-28757325

Abstract: (1) Background: hydrogel scleral buckles (HSB)-related complications can happen decades after implantation, although this material has been retrieved for a long time. Due to its fragile texture, ensuring the complete removal of this material and avoiding complications are challenging. Incomplete removal, iatrogenic complication, recurrent retinal detachment, and infection could occur. (2) Methods: chart review of patients who developed delayed HSB-related complications and received removal of HSB in Taipei Veterans General Hospital from 2004 to 2021. The presenting symptoms, prior diagnosis before referral, clinical findings, image features, surgical technique, operative findings, and outcome were analyzed. Detailed surgical procedure and tips for removal were demonstrated in the study. (3) Results: a total of eleven patients were identified. The presenting symptoms include limitations to extraocular movement (ten eyes, 90.9%), ocular redness (eight eyes, 72.7%), ocular fullness (eight eyes, 72.7%), pain (six eyes, 54.5%), and exposed ocular foreign body (five eyes, 45.5%). Of note, six patients (54.5%) have monocular glaucoma and four of them have intractable high intraocular pressure. All patients underwent surgeries to smoothly remove swollen HSB via transcutaneous or transconjunctival approach. Most symptoms improved after surgery and no cases developed surgical-related complications. (4) Conclusions: although HSB have been off the market for decades, delayed complications are still emerging. Clinicians should remain alert for potential complications for patients with prior HSB surgeries. Early diagnosis and meticulous management can help to safely remove the expanded HSB and reduce the associated complications.

Keywords: hydrogel scleral buckles; retinal detachment; glaucoma; anterior orbitotomy

Citation: Yang, H.-Y.; Yu, W.-K.; Tsai, C.-C. Management of Delayed Complications of Hydrogel Scleral Buckles. *J. Pers. Med.* **2022**, *12*, 629. https://doi.org/10.3390/jpm12040629

Academic Editor: Juan J. Salazar Corral

Received: 5 March 2022
Accepted: 12 April 2022
Published: 14 April 2022

1. Introduction

Hydrogel scleral buckles (HSB) were first introduced as an alternative to silicone buckles for treatment of rhegmatogenous retinal detachment in the 1980s and are thought to possess better softness and elasticity. Softness increase comfort and decreases the rate of scleral erosion and protrusion. Elasticity and mild hydration of hydrogel in the tissue could fill the dead space between the eyeball and the surrounding tissue. In addition, this material in the past was thought to absorb and slowly release antibiotics, further preventing bacterial growth [1]. Complications associated with this material have been reported due to hydrophilic degradation, causing swelling, extrusion, intraocular intrusion, strabismus, infection, bony erosion, and even globe loss [2–9]. It was gradually removed from the market in the 1990s.

Many patients developed HSB-related complications decades after implantation. Some of the cases were initially diagnosed with thyroid-associated orbitopathy, idiopathic orbital fibrosis, orbital tumors, conjunctival cysts, or neurological disorders [4,6,10,11]. Removal of this brittle material was challenging due to its fragile characteristics and the surrounding adhesion to ocular tissue due to fibrosis. The perforation rate during surgery can be as high as 18% [3]. For surgeons who were unfamiliar with this material, incomplete removal, recurrent retinal detachment (up to 29.4%), scleral rupture (up to 17.6%) and other surgical associated complications can occur [3,6].

In this study, we investigated the various clinical manifestations of delayed complications of swollen HSB and shared our surgical techniques to avoid the severe complications associated with manipulations.

2. Materials and Methods

We collected all patients with late HSB complications who underwent surgery for HSB removal in Taipei Veterans General Hospital, a tertiary medical center, from 2004 to 2021. Age, sex, time interval between HSB implantation and removal, presenting symptoms, initial diagnosis, clinical findings, surgical technique, intraoperative findings, and outcome were retrospectively reviewed.

Intraocular pressure (IOP) change was defined as the change in IOP before surgery and one month after surgery. The reason we chose IOP one month postoperatively for comparison is that post-operative tissue edema usually subsides at this time. IOP reduction was defined as a >20% IOP decrease compared to pre-operative IOP.

Location of HSB was categorized into groups according to the relative location to the globe equator in orbit-computed tomography (CT): anterior-located and posterior-located HSB. Anterior-located HSB were defined as more than 50% of HSB coverage anterior to the globe equator in sagittal cut, while the others were defined as posterior-located HSB.

SPSS statistical software version 22.0 (IBM corp., Armonk, NY, USA) was used for data management and statistical analysis. Categorical and paired continuous data were compared using the Fisher's exact test and Wilcoxon Signed ranks test, respectively.

3. Results

In total, we collected eleven cases; Table 1 summarizes the clinical data. One patient had bulbar atrophy before the diagnosis of swollen HSB (Figure 1). Table 2 shows the presenting symptoms of the patients. The symptoms included limitation of extraocular movement (ten eyes, 90.9%), ocular redness (eight eyes, 72.7%), ocular fullness (eight eyes, 72.7%), pain (six eyes, 54.5%), and exposed ocular foreign body (five eyes, 45.5%). Six patients (54.5%) received a correct diagnosis when referred to the medical center. Initial diagnoses other than swollen HSB included eyeball rupture (one eye, 9.1%), and orbital tumor (four eyes, 36.4%). Six patients (54.5%) had glaucoma in the diseased eyes and were treated with at least one combined form or two antiglaucoma medications before orbital HSB swelling was found. Four patients (66.7%) were male, but there was no significance of sex and glaucoma correlation according to Chi-Square analysis ($p = 0.652$). In the six patients with glaucoma, four cases had inadequate IOP control and one case received minimally invasive glaucoma surgery due to uncontrolled IOP one year before HSB removal. New onset of limitations of extraocular movement prompted the doctors to survey for orbital lesions. The typical CT image finding of HSB is a circumferential and homogenous mass around the globe. As this material usually swells when it absorbs water and forms a pseudocapsule around itself, CT images show an isointense signal with a vitreous and non-infiltrative lesion that deforms the eyeball. A scattered, and hyperintense signal, which corresponds to calcific change, can also be found around the swollen HSB (Figure 2). In MRI study, the HSB usually show hypointensity in T1-weighted images and hyperintensity in T2-weighted images, as these materials absorb water (Figure 3). According to the location of HSB in orbital CT, we categorized the HSB coverage into two groups: anterior-located (Figure 4) and posterior-located HSB (Figure 5). In the six patients with glaucoma, four (66.7%) showed posterior-located HSB. However, Fisher's exact test of the relationship between the posterior-located HSB and glaucoma diagnosis did not reach clinical significance ($p = 0.061$).

Table 1. Patient characteristics.

Characteristics	Total $n = 11$
Mean age, years (range)	51 (43–82)
Male, n (%)	7 (63.6)
Mean interval between SB implant and explant, years (range)	20.45 (14–30)
Correct initial diagnosis, n (%)	6 (54.5)
Glaucoma, n (%)	6 (54.5)
Location of SB relative to equator in CT image, n (%)	
Posterior-located	4 (36.4)
Anterior-located	7 (63.6)
Coverage of SB, n (%)	
180 degrees	8 (72.7)
120 degrees	2 (18.2)
90 degrees	1 (9.1)

CT: computed tomography, SB: scleral buckles.

(A) (B) (C)

Figure 1. (**A**) A case of bulbar atrophy after retinal detachment repair with hydrogel scleral buckles. The picture of the left eye showed protrusion of a yellowish, jelly-like material from the orbit, with compression of the eyeball to the medial orbit. (**B**) A yellowish and soft hydrogel scleral buckle dropped from the patient's eye. (**C**) After the foreign body fell out, a hole was left in the superotemporal orbit. The atrophic eyeball was seen, and the surrounding residual hydrogel scleral buckles were noted.

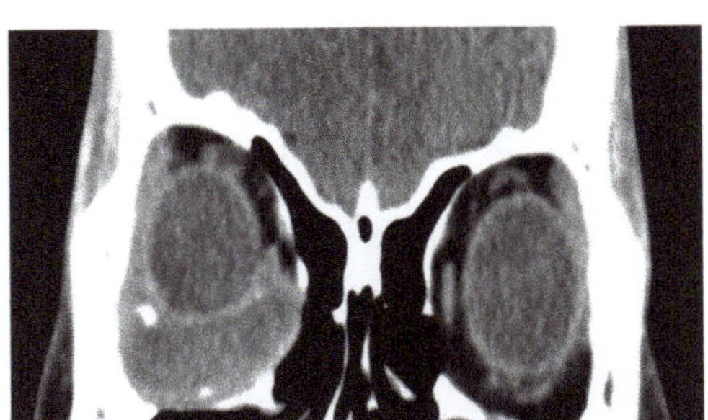

Figure 2. Orbital CT showed a homogenous and hypointense mass with scattered hyperintense spots. This finding was compatible with hydrated hydrogel scleral buckles and dystrophyic calcification.

Table 2. Symptoms before and after hydrogel scleral buckles removal.

Symptoms	Before SB Removal	After SB Removal
Pain, *n* (%)	6 (54.5)	1 (9.1)
Redness, *n* (%)	8 (72.7)	1 (9.1)
Swelling sensation, *n* (%)	8 (72.7)	1 (9.1)
EOM limitation, *n* (%)	10 (90.9)	2 (18.2)
Exposed foreign body, *n* (%)	5 (45.5)	0 (0.0)

EOM: extraocular movement, SB: scleral buckles.

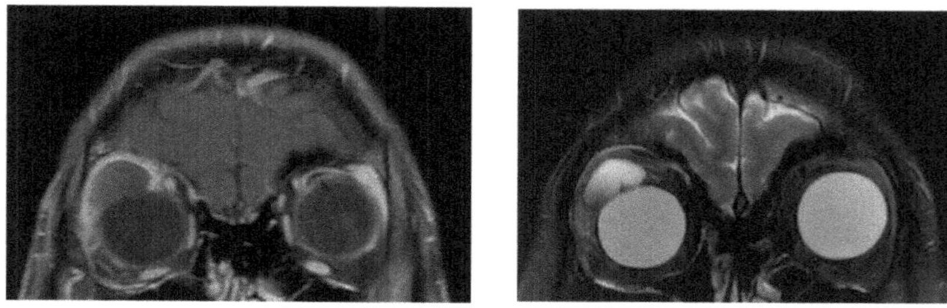

Figure 3. MRI of hydrogel scleral buckles were hypointense in T1 (**left**) and hyperintense in T2 (**right**) images.

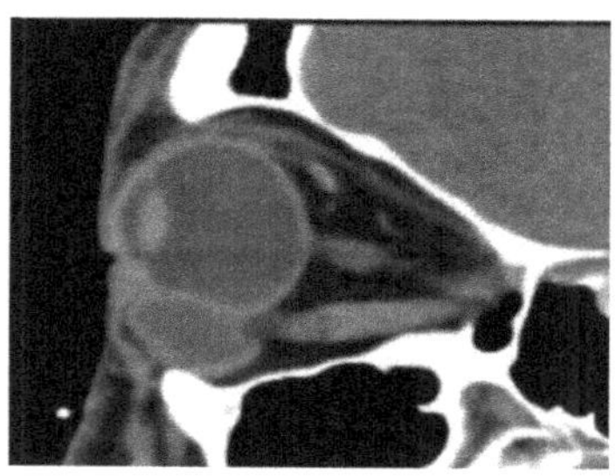

Figure 4. Orbital CT showed anteriorly located hydrogel scleral buckles.

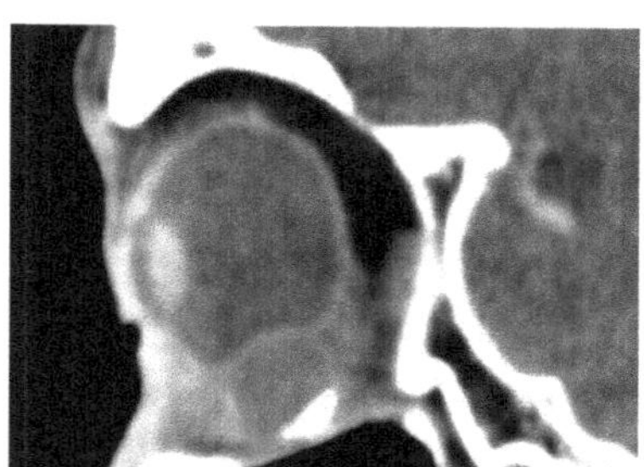

Figure 5. Orbital CT at sagittal view revealed posteriorly located hydrogel scleral buckles.

All cases' HSB were finally removed through one external conjunctival wound if the HSB could be easily approached anteriorly, or through one skin wound if the HSB were posteriorly located. Table 3 summarized the surgical approach of the study cases. Lateral canthotomy was suggested to better expose the surgical field. This jelly-like hydrogel was fragile and can easily be broken into pieces in an attempt to remove it. We used blunt instruments such as muscle hooks or a Desmarres lid retractor to scoop out rather than grab the swelling buckles (Figure 6A,B). A pseudocapsule was usually found during the dissection of orbital space (Figure 6C). Surgeons preferred not to dissect further, as removal of the HSB inside the pseudocapsule was adequate.

Table 3. Surgical approach.

Surgical Approach	Total $n = 11$
Operation under general anesthesia, n (%)	10 (90.9)
Operation under retrobulbar anesthesia, n (%)	1 (9.1)
Transconjunctival approach, n (%)	9 (81.8)
Transcutaneous approach, n (%)	2 (18.2)
Evisceration, n (%)	1 (9.1)

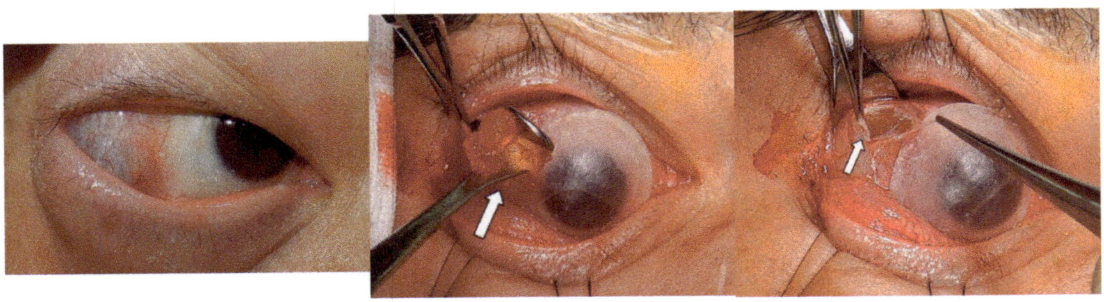

(A) (B) (C)

Figure 6. (**A**) The photo showed protrusion of hydrated hydrogel scleral buckles. (**B**) After lateral canthotomy, removal of the hydrogel scleral buckles with a Desmarres lid retractor (the white arrow) was performed through one conjunctival wound. (**C**) The pseudocapsule (the white arrow) is usually found enclosing the hydrogel scleral buckles.

At the end of surgery, irrigation of the empty space with antibiotics and steroids was performed to prevent infection and decrease inflammation, respectively. The HSB were safely removed in all patients. Evisceration was performed in addition to HSB removal in one case due to severe orbital cellulitis. In the surgical field, no implants were found other than HSB. No intraoperative scleral rupture was noted. There was no endophthalmitis, recurrence of retinal detachment, delayed bleeding or associated complications found in the follow-up period after surgery.

After removal of the HSB, ocular symptoms improved in most cases (Table 2). However, a small residual degree of eye movement limitation was found in two cases. No patients received further strabismus surgery. In the six glaucoma patients, four of them showed IOP improvement one month after HSB removal. There was no change of antiglaucoma medication during this period. However, there was no statistical significance of IOP change before and after HSB removal in the ten cases using Wilcoxon signed ranks test ($p = 0.189$, Table 4).

Table 4. Intraocular pressure before and one month after HSB removal surgery.

Case Number (Medication)	IOP before Surgery (mmHg)	IOP after Surgery (mmHg)
1	Not applicable due to bulbar atrophy before surgery	
2 (G-alphagan, IZBA, carteolol)	27	27
3 (G-cosopt, xalatan)	19	15
4 (G-cosopt)	28	21

Table 4. *Cont.*

Case Number (Medication)	IOP before Surgery (mmHg)	IOP after Surgery (mmHg)
5 (G-lumigan, timolol, trusopt)	35	34
6 (G-duotrav)	8	6
7	15	15
8	14	15
9	13	14
10	12	14
11 (G-alphagan, cosopt, IZBA)	34	19

G: glaucoma case.

4. Discussion

HSB absorb tissue fluids and progressively expand over the decades. The swelling of hydrogel itself has a mass effect on the orbital area, causing compression of the globe. HSB can cause inflammation and fibrosis of the surrounding tissue, and a pseudocapsule usually forms to enclose the hydrogel. If the material distends to a greater extent, it can intrude into the sclera or conjunctiva, or extrude from the orbit [2,4,10]. Delayed HSB complications associated with implant swelling and protrusion are usually found from 54 to 284 months after surgery [3]. The indications to remove the implant include pain, ocular discomfort, ocular inflammation, cosmetic concerns and diplopia [3,6]. In our patient group, the most common symptom was limitations of extraocular movement. The most severe complication of HSB in our study was the marked extension of hydrogel within the orbit, leading to compression, extrusion, and orbital cellulitis in one case. Ocular redness was thought to be caused by an immune reaction to the exposed foreign body. Strabismus was possibly due to myotoxicity, adhesion and scarring [12]. Strabismus can develop after all kinds of scleral buckle implants in around 5–25% of patients in the long term [12,13]. However, HSB-related strabismus has been reported in up to 29% of patients in the previous literature, probably due to its evident mass effect, displacing the ocular content and muscles [3]. As expanded HSB can cause a distinct bulky effect and more severe ocular motility disturbance, the removal of expanded HSB is helpful in reducing the extent of diplopia and the following strabismus surgery.

Incorrect initial diagnosis can lead to incomplete removal and increase iatrogenic complications for unprepared surgeons [4,10]. In our study, four patients (36.4%) were initially diagnosed as having an orbital tumor when they were referred to us. Surgeons who were not familiar with HSB removal may not be able to remove this fragile material safely. The rate of unexpected perforation during surgery ranges from 5.3% to 17.6% [3,6,7]. A comprehensive exam of the retina and supplemental retinal laser may be needed before the operation. Recurrent retinal detachment can occur in from 5.3% to 29.4% of patients after HSB removal [3,6,7]. Endophthalmitis, corneal edema, glaucoma, and scleral necrosis were also reported as complications of removal surgery [6,7].

A proper surgical approach is essential to extract the brittle hydrogel smoothly and avoid disturbing ocular structures. Although the previous literature mentions extensive peritomy, a cryo-assisted approach [14], suction traction [15], or the use of boric acid [16], we use a relatively simple means of removing HSB. An extensive peritomy can result in scleral perforation when dissecting the periocular tissue. Further adhesion and scarring of tissue could ensue. Compression of the globe during manipulation may cause recurrent retinal detachment while the HSB are removed. In contrast, we make a relatively small wound around the skin or fornix and lateral canthotomy to better expose the surgical field. This external approach near the orbital ring can further make a tunnel to the target implant. Usually, the implant is enclosed by a pseudocapsule. The jelly-like material can be extracted with blunt instruments such as a Desmarres lid retractor or muscle hook through the tunnel (Figure 6). At the end of surgery, irrigating the space with steroids and antibiotics may decrease the degree of inflammation and rate of infection.

It is worth mentioning that none of the previous studies mentioned glaucoma status or refractory IOP before removal of the implant. In our study, six patients had glaucoma, and four of them had inadequate IOP control. There are two possible explanations for this. First, these patients had a longer interval between HSB implantation surgery and removal, ranging from 26 to 30 years. Previous large-population studies documented the interval between implantation and removal from 8 to 23 years [3,6,17]. With an extended period in our study, pathological changes in periorbital tissue such as fibrotic change, impaired ocular circulation and ocular compression, can play a role in glaucomatous development. Another explanation is that the relative location and degrees of HSB coverage may influence the development of glaucoma. Four out of the six cases with glaucoma in our study showed HSB coverage posterior to the equator. Compared to the anterior location, posteriorly located HSB cannot release the swelling pressure through protrusion of the orbital exit. Previous studies documented the degrees of HSB coverage but not the relative location to the equator. Roldán-Pallarés et al. followed up 415 cases with hydrogel implantation for up to seven years [17]. More than half of the patients received HSB coverage over 180 degrees, and 45% of the patients had 360-degree encircling HSB coverage. None of these patients reported elevated IOP or glaucoma status, but unspecified orbital fullness was noted in only six cases. These may imply that the relative location compared to the globe equator, not the range of HSB coverage, may be the key factor that influences IOP. The relationship between IOP elevation and the location of swelling HSB must be proven through further studies.

5. Conclusions

Although HSB had not been used for more than 30 years, delayed complications associated with this material still emerged in recent years. Patients could present with various clinical manifestations. A detailed history, examination, and image findings are the keys to correct diagnosis and proper removal of the implant. In our study, an external approach was proved to be a safe way to remove HSB. Blunt instruments are suggested to prevent inadvertent perforation. Early diagnosis and meticulous management are essential to avoid severe sequelae in these patients.

Author Contributions: H.-Y.Y., W.-K.Y. and C.-C.T. helped with data collection. H.-Y.Y. wrote the main manuscript text. All authors have read and agreed to the published version of the manuscript.

Funding: This research received no external funding.

Institutional Review Board Statement: The protocol was approved by the Institutional Review Board of Taipei Veterans General Hospital and supervised by "the Institutional Review Board of Taipei Veterans General Hospital". After a review by Human Research Protection Center, the implementation of the protocol was approved. All the methods were carried out in accordance with relevant guidelines and regulations, and informed consent was obtained from all participants in the study.

Informed Consent Statement: Informed consent was obtained from all subjects involved in the study.

Data Availability Statement: Not applicable.

Conflicts of Interest: The authors declare no conflict of interest.

References

1. Das, T.; Namperumalsamy, P. Scleral buckling with hydrogel implant. *Indian J. Ophthalmol.* **1991**, *39*, 41–43. [PubMed]
2. Bernardino, C.R.; Mihora, L.D.; Fay, A.M.; Rubin, P.A. Orbital complications of hydrogel scleral buckles. *Ophthalmic Plast. Reconstr. Surg.* **2006**, *22*, 206–208. [CrossRef] [PubMed]
3. Crama, N.; Klevering, B.J. The Removal of Hydrogel Explants: An Analysis of 467 Consecutive Cases. *Ophthalmology* **2016**, *123*, 32–38. [CrossRef] [PubMed]
4. Czyz, C.N.; Petrie, T.P.; Everman, K.R.; Cahill, K.V.; Foster, J.A. Hydrogel explant extrusion masquerading as a malignant eyelid lesion. *Ophthalmic Plast. Reconstr. Surg.* **2013**, *29*, e25–e27. [CrossRef] [PubMed]
5. Figueira, E.C.; Francis, I.C.; Wilcsek, G.A. Scleral MIRAgel causing bony orbital erosion. *Orbit* **2007**, *26*, 65–69. [CrossRef] [PubMed]

6. Kearney, J.J.; Lahey, J.M.; Borirakchanyavat, S.; Schwartz, D.M.; Wilson, D.; Tanaka, S.C.; Robins, D. Complications of hydrogel explants used in scleral buckling surgery. *Am. J. Ophthalmol.* **2004**, *137*, 96–100. [CrossRef]
7. Le Rouic, J.F.; Bettembourg, O.; D'Hermies, F.; Azan, F.; Renard, G.; Chauvaud, D. Late swelling and removal of Miragel buckles: A comparison with silicone indentations. *Retina* **2003**, *23*, 641–646. [CrossRef] [PubMed]
8. Roh, M.; Lee, N.G.; Miller, J.B. Complications Assoicated with MIRAgel for Treatment of Retinal Detachment. *Semin. Ophthalmol.* **2018**, *33*, 89–94. [CrossRef] [PubMed]
9. Rubinstein, T.J.; Choudhary, M.M.; Modi, Y.S.; Ehlers, J.P.; Perry, J.D. Globe Loss from Intraocular Invasion of MIRAgel Scleral Buckle Components. *Ophthalmic Plast. Reconstr. Surg.* **2016**, *32*, 329–332. [CrossRef] [PubMed]
10. Shields, C.L.; Demirci, H.; Marr, B.P.; Mashayekhi, A.; Materin, M.A.; Shields, J.A. Expanding MIRAgel scleral buckle simulating an orbital tumor in four cases. *Ophthalmic Plast. Reconstr. Surg.* **2005**, *21*, 32–38. [CrossRef] [PubMed]
11. Sadaka, A.; Ortiz, J.; Berry, S.; Lee, A.G.; Li, H.K.; Divatia, M.; Malik, A. Late proptosis and ophthalmoplegia from hydrogel scleral buckle. *Can. J. Ophthalmol. J. Can. d'ophtalmologie* **2018**, *53*, e190–e193. [CrossRef] [PubMed]
12. Farr, A.K.; Guyton, D.L. Strabismus after retinal detachment surgery. *Curr. Opin. Ophthalmol.* **2000**, *11*, 207–210. [CrossRef] [PubMed]
13. Rabinowitz, R.; Velez, F.G.; Pineles, S.L. Risk factors influencing the outcome of strabismus surgery following retinal detachment surgery with scleral buckle. *J. Am. Assoc. Pediatric Ophthalmol. Strabismus* **2013**, *17*, 594–597. [CrossRef] [PubMed]
14. Le Rouic, J.F.; Bejjani, R.; Azan, F.; Bettembourg, O.; Renard, G.; Chauvaud, D. Cryoextraction of episcleral Miragel® buckle elements: A new technique to reduce fragmentation. *Ophthalmic Surg. Lasers* **2002**, *33*, 237–239. [CrossRef] [PubMed]
15. Richards, A.L.; Meyer, D.R. Late complications of hydrogel scleral buckle implants and a technique for effective removal. *Ophthalmic Plast. Reconstr. Surg.* **2012**, *28*, 455–458. [CrossRef] [PubMed]
16. Roldán-Pallarés, M.; Avila, M.I.; Refojo, M.F. The use of boric acid solution to help in the removal of biodegraded Miragel episcleral buckles. *Retina* **2005**, *25*, 90–93. [CrossRef] [PubMed]
17. Roldán-Pallarés, M.; Hernández-Montero, J.; Llanes, F.; Fernández-Rubio, J.E.; Ortega, F. MIRAgel: Hydrolytic degradation and long-term observations. *Arch. Ophthalmol.* **2007**, *125*, 511–514. [CrossRef] [PubMed]

MDPI
St. Alban-Anlage 66
4052 Basel
Switzerland
Tel. +41 61 683 77 34
Fax +41 61 302 89 18
www.mdpi.com

Journal of Personalized Medicine Editorial Office
E-mail: jpm@mdpi.com
www.mdpi.com/journal/jpm

www.ingramcontent.com/pod-product-compliance
Lightning Source LLC
Chambersburg PA
CBHW061630210926
51850CB00009B/2497
* 9 7 8 3 0 3 6 5 8 4 0 3 4 *